D0520260

"Information through Innovation"

DATA COMMUNICATIONS AND COMPUTER NETWORKS

AN OSI FRAMEWORK

CURT M. WHITE

INDIANA UNIVERSITY-PURDUE UNIVERSITY FT. WAYNE

boyd & fraser publishing company

I(T)P   An International Thomson Publishing Company

Danvers • Albany • Bonn • Boston • Cincinnati • Detroit • London • Madrid • Melbourne
Mexico City • New York • Paris • San Francisco • Singapore • Tokyo • Toronto • Washington

Executive Editor: James H. Edwards
Project Manager: Christopher T. Doran
Production Editor: Barbara J. Worth
Manufacturing Coordinator: Tracy Megison
Marketing Coordinator: Daphne J. Meyers
Production Services: Stacey Sawyer,
Sawyer & Williams

Composition: Fog Press
Illustration: Fog Press
Interior Design: John Edeen
Cover Design: Mike Fender Design
Cover Photo: E. Salem Krieger/The Image Bank

bf   © 1995 by boyd & fraser publishing company
     A Division of International Thomson Publishing Inc.

I(T)P   The ITP™ logo is a trademark under license.

Printed in the United States of America

For more information, contact boyd & fraser publishing company:

boyd & fraser publishing company
One Corporate Place • Ferncroft Village
Danvers, Massachusetts 01923, USA

International Thomson Publishing Europe
Berkshire House 168-173
High Holborn
London, WC1V 7AA, England

Thomas Nelson Australia
102 Dodds Street
South Melbourne 3205
Victoria, Australia

Nelson Canada
1120 Birchmont Road
Scarborough, Ontario
Canada M1K 5G4

International Thomson Editores
Campos Eliseos 385, Piso 7
Col. Polanco
11560 Mexico D.F. Mexico

International Thomson Publishing GmbH
Königwinterer Strasse 418
53227 Bonn, Germany

International Thomson Publishing Asia
221 Henderson Road
#05-10 Henderson Building
Singapore 0315

International Thomson Publishing Japan
Hirakawacho Kyowa Building, 3F
2-2-1 Hirakawacho
Chiyoda-ku, Tokyo 102, Japan

1 2 3 4 5 6 7 8 9 10   MT   8 7 6 5 4

Library of Congress Cataloging-in-Publication Data
White, Curt M.
    Data communications and computer networks : an OSI approach / Curt M. White
        p.   cm.
    Includes index.
    ISBN 0-7895-0053-1
    1. Computer networks.   2. OSI (Computer network standard)   3. Data transmission systems.   I. Title.
TK5105.5.W467 1995   94-37161
004.6--dc20   CIP

Photo credits: page 9 Chrysler Corporation; page 12 U.S. Sprint; page 58 IBM; page 63 (left) Hayes; page 63 (right) Multi-Tech Systems.

CONTENTS

3 PHYSICAL LAYER: PART TWO   51

4 THE DATA LINK LAYER   79

5 THE NETWORK LAYER   107

6 LOCAL AREA NETWORKS   127

7 WIDE AREA NETWORKS   163

8 SATELLITE AND RADIO BROADCAST NETWORKS   187

9   TRANSPORT AND SESSION LAYERS   203

10   PRESENTATION AND APPLICATION LAYERS   221

PREFACE

Capturing the field of data communications and computer networks in writing is similar to hitting a target moving at the speed of light. New techniques and technologies emerge so quickly it is nearly impossible to write a textbook that incorporates up-to-the-minute advancements.

Intended Audience

This text is designed for use in introductory undergraduate computer information systems and computer science courses. Textbooks for these courses typically fall into one of two categories: a fairly technical, engineering-like approach to the subject; or a business-like approach that introduces much terminology in a breadth-first strategy. Topics herein are presented with both sufficient technical explanation and in a framework that allows the reader to apply these topics either to more advanced concepts or to the real world. The more difficult topics such as the OSI model, IEEE 802, TCP/IP, wireless networks, X.25, and SNA are presented in a clear, readable fashion, yet with sufficient technical details to facilitate understanding.

Content Organization

The outline of the text follows the seven layers of the OSI model. After an introduction to basic terminology, Chapter 2 and Chapter 3 discuss the physical layer of the OSI model, including media types and characteristics, signals, data codes, types of errors, multiplexing, terminals, modems, and interface standards. Chapter 4 presents the data link layer of the OSI model by discussing error detection, asynchronous transmission, BISYNC, SDLC and HDLC synchronous transmission, and the medium access control layer. Chapter 5 introduces the network layer with an important discussion of network terminology, followed by routing techniques and network congestion.

Chapter 6 presents the topic of local area networks, including network topologies, medium access control protocols, sample networks, and a solid introduction to internetworking local area networks. Chapter 7 introduces wide area networks, including X.25 networks, IBM's SNA, and the popular Internet and the Internet Protocol. Chapter 8 discusses the basics of satellite and radio broadcast networks, including the quickly growing cellular and wireless networks.

Chapter 9 presents the transport and session layers of the OSI model, along with a solid introduction to the TCP portion of the TCP/IP Internet protocols. Chapter 10 introduces the presentation and application layers, including a presentation of the Abstract Syntax Notation, compression techniques, encryption and decryption, electronic mail, file transfers, and remote logins. Chapter 11 presents some of the emerging technologies such as ISDN and broadband-ISDN, asynchronous transfer mode, wireless technology, and the "information superhighway."

The ongoing case study presented at the end of every chapter allows the reader to incorporate newly learned skills into a real-life situation. It is strongly recommended that both student and instructor spend time reading, presenting, and discussing the problems and possible solutions. One should note that there may be other solutions that are as acceptable, if not better, than the solution given in the text.

Acknowledgments

I would like to thank the people at International Thomson Publishing, including Kathy Shields and Tamara Huggins, at Wadsworth, and James Edwards, Christopher Doran, and Barbara Worth, at boyd & fraser, for their support and encouragement. I would also like to thank the reviewers who spent a considerable amount of time reading the manuscript and providing helpful suggestions: C. T. Cadenhead, Richland College; David Doss, Illinois State University; and Patricia A. McQuaid, Auburn University.

Finally, I would like to thank my wife, Kathleen, for her never-failing support and encouragement. Her wisdom and patience gave me the strength to complete this project.

Curt M. White

CHAPTER 1

INTRODUCTION TO DATA COMMUNICATIONS AND COMPUTER NETWORKS

INTRODUCTION

The world of data communications is a surprisingly vast and increasingly significant field of study. Once considered primarily the domain of engineers and technicians, data communications now involves computer programmers (both scientific and commercial), managers, system designers, and home computer users. Even for the average person on the street, rarely does a day pass that he or she does not use some form of data communications because the basic concepts of this field are applied so widely in our daily lives.

The telephone system is one of the oldest and most common vehicles for data communications. At one time a relatively simple, analog medium for carrying voice conversations, it has evolved into a complex, digital system capable of transmitting voice, video, and music over advanced glass wires. As the digital age of electronics continues to shape the U.S. telephone system, more and more advanced features such as caller identification, multiple party conferencing, and the blocking of designated incoming phone numbers are being added.

On the business front, banking systems use more data communications every day. Twenty-four-hour automatic teller machines (ATMs) can perform most of the day-to-day functions of withdrawals, deposits, and transfers from locations worldwide. The codes and mechanisms for transferring information from the ATM to the host bank and back again are used throughout the data communications field.

Audio and video systems continue to push the known thresholds of digital communications. The techniques of digital television, the compact disc player, and cable and satellite television systems borrow many concepts from the field of data communications.

The wire that currently delivers cable television to many homes will soon provide access to so many worldwide services as to boggle the imagination.

Even our homes and automobiles have begun to contain data communications systems. More and more residences have advanced electronic burglar alarms or remote controls for operating lights and appliances. Many of these conveniences use communication codes and data transmission procedures. The automobile has become, to many, an office on wheels, containing mobile telephones, facsimile (fax) machines, and portable computers.

As we see the growing importance of data communications, we realize we cannot leave this area of study to the technicians alone. All of us, particularly computer science and information systems students, need to understand the basic concepts. Welcome to the amazing world of data communications!

A GENERIC VIEW OF DATA COMMUNICATIONS

If we could examine the field of data communications as a simplified box diagram, we might see something like Figure 1.1. (Note: terms introduced in this section will be explained in further detail in the chapters that follow.) Beginning at the top left, a **User/device** produces some form of information and presents it to an **input device**. This

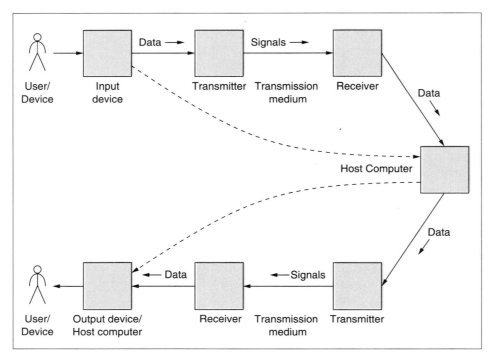

FIGURE 1.1   Simplified box diagram of data communications

user/device could be a person speaking into a telephone, someone typing on the keys of a computer terminal, or a blast furnace melting steel.

The input device accepts this information and produces some form of data. The input device could be a telephone, a computer terminal, a microcomputer, or a temperature sensor.

To be processed, the data is transferred to some form of information processor or computer. This transfer occurs over a **transmission medium**, such as a conventional phone line, a coaxial cable, a radio wave, or a fiber-optic cable.

To be transmitted over a particular transmission medium, data often must be converted to a signal. The **transmitter** performs the data-to-signal conversion. The type of signal created is chosen to take greatest advantage of the existing medium. Modems (modulator/demodulator) and codecs (coder/decoder) are examples of devices that input data and output a specially prepared signal. Just before the signal reaches the host computer, another modem or codec (**receiver**) converts the signal back to the original data form, which is then delivered to the **host computer** for processing.

This may be the end of the line for the original information. Many times, however, new information or a response is returned to the source user/device or to another user/device. In this case, the information makes a return trip through transmitter, transmission medium, receiver, and **output device**.

If the distance between the input device and the host computer is relatively short, there may not be a need to convert the data to a signal and back again. In this case, the transmitter and receiver are not used, as shown by the arcs bypassing these two devices in Figure 1.1. If the output device is a second host computer, we have the beginnings of a computer network—a collection of interconnected processing units that provides a data communications service.

THE COMPLETE GENERIC MODEL

To the basic generic model outlined earlier we can add a few more features commonly found in real-life data communications systems, thus creating a more complete model.

How does the input device connect to the transmitter? What does the connector or plug look like? How many pins? What are the voltage levels? If the signal on Pin 3 becomes active, is that followed by an active signal on Pin 6? This is the topic of **interfacing** and will be discussed in detail in Chapter 3.

What happens if the data/signal is corrupted as it is transmitted from one point to another? What do we have to do to the original data to enable us to discover that an error has occurred and that we are not receiving the same data that was transmitted? If an error is detected, how do we correct it (if at all)? This is **error detection and correction** and is covered in Chapter 2.

What code is used to represent the data between the input device and the transmitter? That is, what sequence of 1s and 0s is used to represent the letter A, the letter B, and so on? Do we use such common codes as **ASCII** (American National Standard Code for Information Interchange) or **EBCDIC** (Extended Binary Coded Decimal Interchange Code), or do we use another? Are some codes better than others in certain situations? Chapter 2 discusses data codes.

What if the data needs to be secure from wiretapping and illegal access as it is being transmitted? This is the topic of **encryption** and **decryption**, covered in Chapter 10.

What if one host computer wishes to communicate with another host computer? What if the distance is short? What if the distance is across the country? These are the topics of **networking** and will be covered in more detail in Chapters 5, 6, 7, and 8.

At this point you should begin to realize just how large and involved the field of data communications actually is. Entire textbooks have been devoted to the individual topics above. At the conclusion of this text, we will have introduced all the parts of the generic model.

U.S. Telephone System Structure

Perhaps the most familiar communications device is the telephone. The U.S. telephone system of today is an enormously complex network compared to the phone system of just 40 years ago. As more and more phones are added daily, the system must expand to

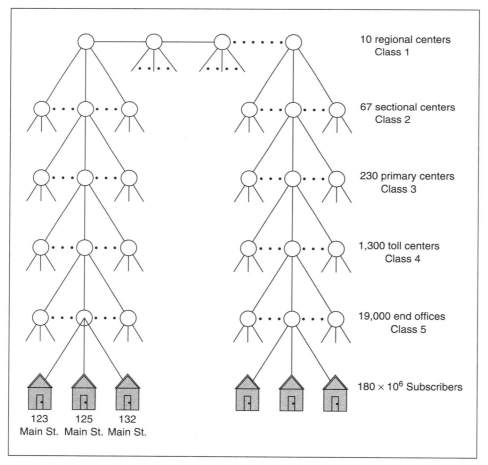

FIGURE 1.2 Hierarchical telephone system in the United States

accommodate them. In fact, the number of telephones is growing so rapidly that in 1993 all the available area codes with a middle digit of 0 or 1 were exhausted. Plans are currently under way to revamp the entire system so that it will support area codes with other middle digits.

In order to handle the vast numbers of telephones, the system was designed as a hierarchical network as shown in Figure 1.2. The sharing of upper-level phone lines is based on the assumption that not everyone will use his or her telephone at the exact same moment. When everyone does want to talk at once, a telephone traffic jam of sorts results. You may have experienced this if you have ever tried to place a long distance call on a busy holiday and were not able to secure a long distance circuit. One solution the telephone companies might use to solve this problem—although not a very practical one—would be to have every telephone in the United States directly wired to every other telephone. The telephone in your house would have a direct wire from you to Aunt Millie, a direct wire from you to your place of work, a direct wire from you to the grocery store, and so on (Figure 1.3). Could you imagine 600 million pairs of wires entering your house? That would be a pretty hefty cable!

Instead, the wire that leaves your house travels to the local telephone company switching office, or Class 5 office (Figure 1.4). All telephones with the same first-three-digit telephone number (prefix, not area code) go to the same Class 5 office. If you call someone with the same prefix as yours, your telephone call enters the local Class 5 office and is simply connected (or switched) to the outgoing line of the telephone number you are dialing. If you are placing a call with a prefix different from yours, the call may take different routes depending on the distance involved. For example, under certain circumstances, your call may simply pass from your Class 5 office to the next Class 5 office. Or your call may go from your Class 5 office up to a Class 4 office and then to the destination Class 5 office. The greater the distance of your telephone call, the farther up the hierarchical tree it may progress.

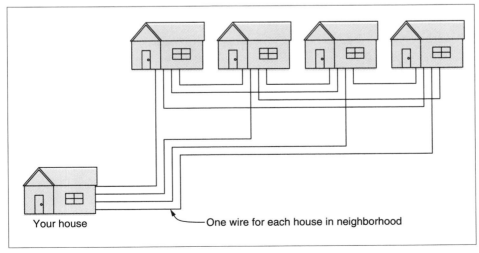

Your house   One wire for each house in neighborhood

FIGURE 1.3   Wiring connections if your house were wired to all other houses in your neighborhood

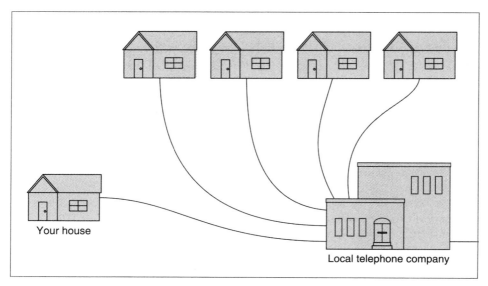

FIGURE 1.4 Wiring connections when all houses in your neighborhood connect only to the local telephone company

Each state is assigned one or more **Local Access and Transport Areas** (**LATAs**). As long as you call someone within your LATA, the telephone call can be handled by your local telephone company. Any time your call leaves your assigned LATA and enters another, your call is considered a long distance call spanning multiple LATAs. Local phone companies cannot make these long distance calls but must pass them on to a long distance telephone company such as U.S. Sprint, MCI, or AT&T. For information on the division of your area into LATAs, consult the front of your telephone directory.

AT&T DIVESTITURE

The organization of a state into Local Access and Transport Areas was one of the results of a significant court case between the U.S. government and AT&T. Since the 1940s, various organizations have fought the monopoly that AT&T has held on the tele-communications industry in the United States with one court case after another chipping away at its grip. The Modified Final Judgment of 1984 was the most far-reaching of these court decisions and caused the divestiture of AT&T. Under this judgment, AT&T was forced to rid itself of all telephone companies that provided local service but was allowed to keep all the long distance lines as well as Bell Laboratories. At the time, AT&T consisted of 23 Bell Operating Companies (BOCs) that provided local phone service across the country. There were also as many as 14,000 other local tele-phone companies not owned by AT&T that typically served very limited areas. With the divestiture, the 23 BOCs were separated from AT&T and, in order to survive, reor-ganized into seven Regional Bell Operating Companies (RBOCs) (Figure 1.5). These seven RBOCs then competed with the remaining 14,000-plus local telephone compa-nies to provide local service to each home and business in the country. Because of the

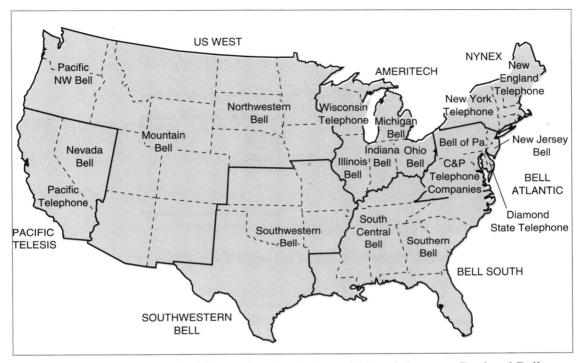

FIGURE 1.5   Map of the United States showing the boundaries of the seven Regional Bell
Operating Companies (RBOCs)

power and size of the seven RBOCs, the 14,000-plus local telephone companies were
reduced to approximately 1400.

While the 1984 divestiture court case may not seem major at this point, it was
perhaps the most important case in our history regarding telephone systems (telecommunications) and data communications. By breaking up AT&T into smaller pieces, the
court reduced the threat of monopoly and opened the door for the Regional Bell Operating Companies, other telephone companies, cable television companies, and many
more private companies to provide a wide range of communication services to homes
and offices. Buzzwords such as *information superhighway* try to describe the vast number of information services that may soon be available to everyone via the telephone
and cable television service. The next section sketches out the broad range of these
applications.

APPLICATIONS OF DATA COMMUNICATIONS

At the beginning of this chapter we enumerated a few of the many application areas of
data communications. Looking at a more complete list can be quite informative, but
rather than setting out all the different types of jobs and services that use some sort of
data communications (the list would be enormous), we can see how extensive the uses
are by examining the general application areas.

■ **Computer terminal to mainframe computer connection**—Many business systems employ a terminal-to-mainframe connection. These systems include inquiry/response applications, interactive applications, and data entry applications, such as airline reservation systems (Figure 1.6).

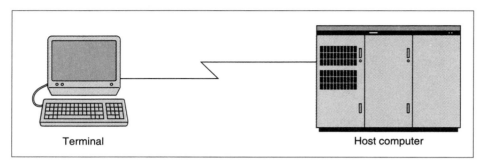

FIGURE 1.6   Simple example of a computer terminal connected to a mainfram computer

■ **Microcomputer to mainframe computer connections**—It is quite common for microcomputer users to connect their machines to mainframe computers, thus being able to draw on the best of both. For example, many companies provide employees with desktop microcomputers. If they wish to access the mainframe computer, these users can download information from the mainframe to the microcomputer. Also, many home microcomputer users dial in to remote mainframe computers via the telephone. Bulletin boards and online information services are very popular with home users (Figure 1.7).

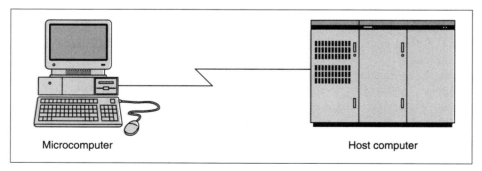

FIGURE 1.7   Example of a microcomputer connected to a mainframe computer via a telephone line

■ **Sensor-based systems**—Manufacturing applications often connect sensors to data gathering computers for industry management, process control, and real-time systems. For example, automobile assembly lines use sensor-based systems extensively for controlling the assembly of vehicles and for the safety of human workers (Figure 1.8).

FIGURE 1.8   Assembly line that uses sensors connected to a computer system

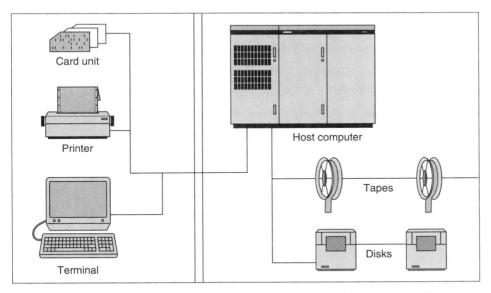

FIGURE 1.9   Simple example of a batch computer operation connected to a
mainframe computer site

- **Batch systems**—Commercial corporations often use batch-oriented systems for running large, data-intensive applications, such as the processing of claims an insurance company might perform. Typically, batch systems process thousands of records of data during off-peak hours when computer access is cheaper or faster than during the normal workday (Figure 1.9).

- **Computer to computer connections**—This application enters the realm of computer networking and can involve a wide range of communication protocols and standards (Figure 1.10).
- **Network to network connections**—Often it is necessary to connect two separate networks. If the two are similar, simple devices can be used for the interconnection. If they are dissimilar, complex devices and protocols are necessary to allow the two networks to share information (Figure 1.11).

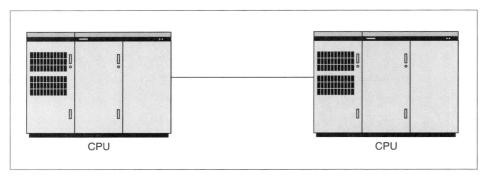

FIGURE 1.10   Two mainframe computers directly connected

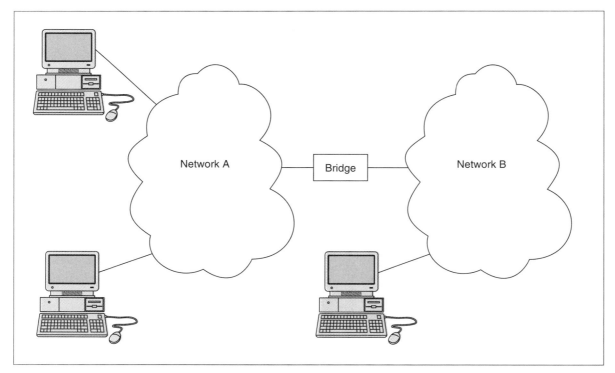

FIGURE 1.11   Simple example of two computer networks bridged together

■ **Satellite and radio network systems**—Quick transfer of vast amounts of data can be accomplished with satellite and radio networks. These could include microwave telephone transmissions, transworld television transmissions, home satellite television, and emergency warning systems. Such systems employ unique and complex protocols to accomplish efficient data transfers (Figure 1.12).

■ **Mobile network systems**—These systems represent a rapidly growing area that includes cellular car phones, emergency dispatching systems, pager systems, and automobile satellite directional systems (Figure 1.13).

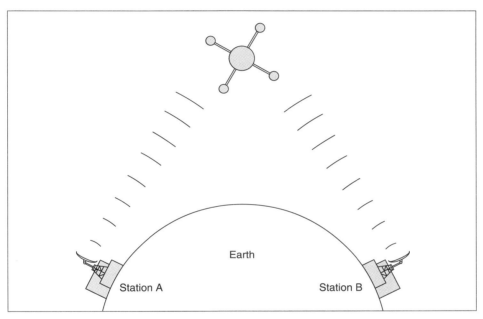

FIGURE 1.12   Example of a satellite system with two earth stations transmitting data

FIGURE 1.13   Driver of an automobile using a mobile telephone network

- **Home control systems**—Modern homes may now be equipped with automatic burglar alarm systems, remote entry systems, remote heating and cooling controls, and remote light and appliance controls. Each of these applications requires extensive communication procedures, and companies are scrambling to be the first to introduce such products and standards.
- **Client-server systems**—An area of computer communications that has grown very popular in the last several years is the client-server system. Unlike a peer-to-peer connection, in which both ends of the communication system are equal in capability, the client-server system involves a workstation (client) that requests information from a server. The local area network (discussed in detail in Chapter 6) is one example of a client-server system that involves one or more workstations requesting information from a network file server.

All the applications described in this section involve the transfer of data from one point to another. If this transfer is to be accomplished successfully, it requires a well-ordered passageway; in such a structure, each piece plays a specific part to ensure the integrity and safe arrival of the data at its destination. One such structure—in this case the Open Systems Interconnection Reference Model—is discussed next.

THE OPEN SYSTEMS INTERCONNECTION (OSI) REFERENCE MODEL

Most large organizations that produce some type of product have a division of labor. Secretaries do the paperwork; accountants keep the books; laborers perform the manual duties; scientists design products; engineers test them; managers control operations. Rarely is one person capable of performing all these duties. Large software applications operate the same way. Different procedures perform different tasks. The whole would not function without proper operation of each of its parts. Communications software is no exception. As the size of the applications grows, the need for a division of labor becomes increasingly important.

A communications application—such as an electronic mail (E-mail) system that accepts the message "Bill, how about lunch? Sally"—has many parts. To begin, the E-mail "application worker" prompts the user to enter a message and an intended receiver. The application worker creates the appropriate packet with message contents and addresses and sends it to a "presentation worker." The presentation worker examines it to determine whether encryption or data compression is required and if so, performs the necessary functions. The packet of information is then passed to the "session worker" that is responsible for adding session information and backup synchronization points in case of failures. Next the packet goes to a "transport worker" that is responsible for providing overall transport integrity. The "network worker" receives the packet next and may add routing information so that it may find its way through the network (if there is one). Next to receive the packet is the "data link worker" that inserts error-checking information and prepares the packet for transmission. Finally the "physical worker" receives the packet and transmits it over the chosen medium.

This is basically the scheme adopted by the International Standards Organization (ISO) when its members created the **Open Systems Interconnection (OSI)** Reference Model. As shown in Figure 1.14, the model consists of seven layers. Note that the layers do not specify precise protocols or exact services but rather define a *model* for the functions that should be performed.

The physical layer handles the transmission of raw bits over a communications channel, including voltage levels and bit durations. Pin configurations, connector dimensions, and other mechanical issues are addressed here.

The data link layer is responsible for taking the raw data and transforming it into a cohesive unit called a frame. The frame should appear error free to the next layer, the network layer. Therefore, the data link layer inserts an error detection protocol before transmitting a packet and performs the error checking on receiving a packet. If there is an error, the data link layer is responsible for informing the sender of the error. It also must make sure the transmitter does not overrun the receiver with too much data (flow control).

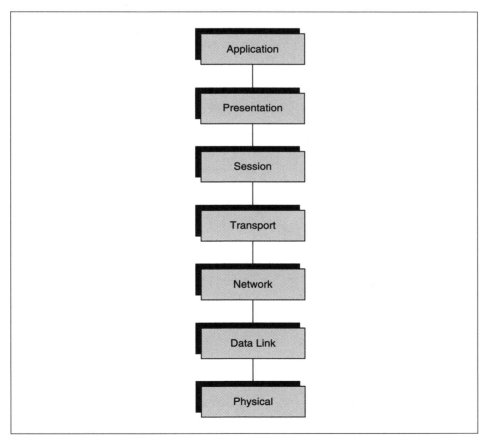

FIGURE 1.14   Open Systems Interconnection (OSI) seven-layer model

The network layer sends the packet of data from node to node within the subnet of the network. To accomplish this, routing information is determined and applied to each packet or group of packets. The network layer also ensures that the network does not become saturated at any one point (congestion control). In networks that use a broadcast distribution scheme, in which the transmitted data is sent to all other stations, the network layer may be very simple or omitted altogether.

The first three layers deal with passing the data packet from node to node in the network/system. The fourth layer, the transport layer, is the first one concerned with an error-free end-to-end flow of data. To ensure that the data arrives error-free at the destination, the transport layer must be able to work across all kinds of reliable and unreliable networks. Comparing it to a manager who must cover for both reliable and unreliable workers helps you see how large and important the transport layer is.

The session layer is responsible for establishing sessions between users and also for handling the service of token management, which controls who talks when during the current session. Additionally, the session layer establishes synchronization points or backup points in case of errors or failures. For example, while transmitting a large document, the session layer may insert a synchronization point at the end of each chapter. If an error occurs during the transmission, both sender and receiver can back up to the last synchronization point (beginning of a chapter) and start retransmission from there.

The presentation layer performs a series of miscellaneous functions necessary for presenting the data packet properly to the sender or receiver. For example, the presentation layer can perform ASCII to EBCDIC conversions, encryption/decryption, and compression functions.

The final layer is the application layer. While many kinds of applications use computer networks, certain of these have received widespread use and thus have been standardized. Applications such as electronic mail, file transfer systems, remote job entry systems, and directory services have been standardized (or at least attempts at this have been tried).

THE INTERNET MODEL

The OSI model is only one model that is used to create large communication systems software. A second model that continues to gather many supporters is the Internet model which incorporates the very popular TCP/IP protocols. The Internet model began as a creation of the U.S. military, but has grown to include a large community of non-military users, including scientists, academicians, businesses, and the government. Figure 1.15 demonstrates how the Internet model compares to the OSI model.

The Interface layer of the Internet model is equivalent to the physical and data link layers of the OSI model. The Internet layer is roughly equivalent to OSI's network layer, but includes the very popular Internet Protocol (IP) for transferring data between networks. The Transport layers of the two models are roughly equivalent. However, the Internet model's Transport layer includes the very popular Transmission Control Protocol (TCP). The final layer of the Internet model, the Application layer, contains the network applications for which one uses a network. The Internet model applications include some well-used programs, including:

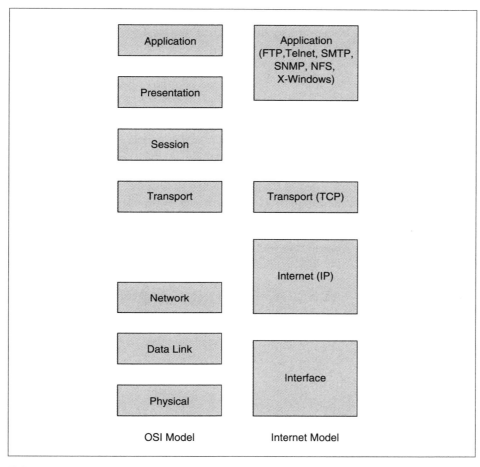

FIGURE 1.15   Comparison of Internet model to the OSI model

FTP (File Transfer Protocol)—for transferring files from one computer system to another.

Telnet—to allow a remote user to login to another computer system

SMTP (Simple Mail Transfer Protocol)—to allow users to send and receive electronic mail

SNMP (Simple Network Management Protocol)—a protocol which allows all the numerous forms of computer networks within the Internet to be viewed as a simple network

NFS (Network File System)—designed by Sun Microsystems to allow access to UNIX file systems

X-Windows—set of protocols that define a graphic window system for networks.

While the Internet model has experienced more widespread use than the OSI model, we will follow the OSI model basically as a tool to present the various parts of computer network software.

This chapter thus far has introduced a number of concepts and applications. Left in the abstract, these can seem quite removed from everyday life. To make them real, we have created a case study centered on the XYZ University where we will apply the new concepts we learn in each chapter. The basic structure of the university is described next.

INTRODUCTION OF CASE STUDY

The XYZ University is an imaginary university located somewhere in the remote regions of your state. The campus is small and contains only three buildings (Figure 1.16). Building A contains the mainframe computer, which is capable of supporting approximately 50 computer terminals. Also located in Building A is the staff of the computing center, who support the mainframe computer and all associated software and hardware.

Building B contains most faculty offices, including those of the computer faculty. Buildings A and B are separated by approximately 100 feet with a small tunnel running between the buildings. The library is the third building and is approximately 100 feet from Building B, to which it is also connected by a small tunnel.

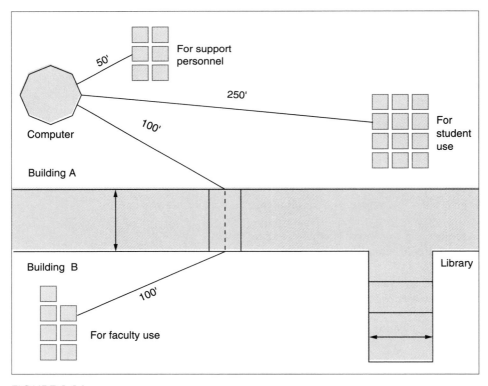

FIGURE 1.16   Basic campus layout of the XYZ University

The computing center wishes to support batch and online transactions with roughly 30 terminals and approximately five dial-up ports. A number of terminals will be needed in Building A for the support personnel of the computing center staff. These terminals will be located near the mainframe computer. The computing center staff has also decided to locate a number of terminals in Building A for students who are taking computer courses and need access to the mainframe computer.

Building B will house the computer faculty, and thus a number of computer terminals will be located in their offices for their own use and for assisting the students. The librarians have not yet decided how they will use the mainframe computer in library operations, so we will defer their computer needs until a later time.

At the completion of each chapter, the computer center staff will consider and decide how best to implement the newly introduced technology. Money, as always, is an issue, and less expensive ways of performing operations will always be sought, as long as they do not result in a significant degradation of services. The staff of the hypothetical computing center will be guided by the important old axiom: time is money. If a particular implementation is initially less expensive but manages data transmission less efficiently or more slowly, the computing center staff will consider the longevity of the proposed solution.

SUMMARY

Data communications is a vast and significant field of study. Once considered the province primarily of engineers and technicians, it now involves computer programmers, managers, system designers, and home computer users. Many common services and products that we use every day employ data communications in one way or another. Telephones, banking systems, cable television, audio and video systems, home burglar alarm systems, and mobile telephones are just a few examples.

Perhaps the most common data communications device is the telephone. The basic structure of the U.S. telephone system is a hierarchical network of class offices. The distinction between local calls and long distance calls rests with the concept of the Local Access and Transport Area (LATA). The divestiture of AT&T plus the growth of digital communications is leading to an information superhighway that will someday interconnect homes and offices to a vast number of online computer resources.

The application areas of data communications and computer networks cover a broad spectrum with too many uses to enumerate here. We chose instead to create a general list of application areas, among which are batch computer systems, sensor-based computer systems, microcomputer to mainframe links, mainframe computer to mainframe computer links, network to network links, and satellite and radio networks.

To standardize the design of communication systems, the International Standards Organization (ISO) created the Open Systems Interconnection (OSI) model. The OSI model is based on seven layers beginning with the physical layer and ending with the application layer. While this is only one model among many, the OSI model is growing in importance and roughly forms the basic outline for the remainder of the text.

EXERCISES

1. Create a list of all the things you do in an average day that use data communications.
2. If you could design your own home, what kinds of data communications/labor-saving devices would you incorporate?
3. Using your local telephone directory, determine the boundaries of the LATA in which you live and list the major cities in it.
4. Summarize each layer of the OSI model in one, grammatically correct sentence for every layer.
5. Using the generic model in Figure 1.1
   a. list five types of input devices
   b. list five types of output devices
6. a. How many different phone numbers can exist within one area code?
   b. If we are limited to area codes with a 0 or 1 as the second digit, how many area codes are possible? How many different phone numbers?
   c. If we allow numbers other than 0 or 1 as the second digit in the area code, how many area codes are possible? How many different phone numbers?

CHAPTER 2

THE PHYSICAL LAYER: PART ONE

INTRODUCTION

As we begin to learn the more technical aspects of data communications, we will start at the ground level and work our way up. Relating our coverage of the material to the Open Systems Interconnection (OSI) seven-layer model, we begin our discussion with layer one, the physical layer. You will recall that the physical layer is concerned with decisions such as choice of media, voltage levels, plug configurations, and the electronic boxes that perform various conversions. We examine each of these topics, and more, in the following two chapters.

MEDIA

The world of data communications would not exist if there were no medium by which to transfer data. While there have been few recent and unique additions to the list of media types—fiber optic cable being the newest member of the family—improvements in design and technology are continuously increasing the capabilities of existing media. Since this textbook was published, processor speeds and technological advancements have increased even further. Undoubtedly, as you read this paragraph, someone somewhere is designing new materials and building new equipment that is better than what currently exists. We must keep this evolution in mind as we examine the basic characteristics of the major types of media. The values given for transmission speeds and equipment costs are, at most, relative values, to be used mainly for comparing one technology to another.

The family of transmission media can be divided into two classes: the conducted (hardwire) media, and the radiated (softwire) media. Conducted media are physical wires or cables that pass electromagnetic or light signals. Radiated media consist of techniques of transmitting radio waves through the air and space.

Conducted Media

Twisted Pair Cable

The oldest, simplest, and most common type of conducted media is **twisted pair cable**. Its name describes it: a pair of wires, one twisted around the other, as shown in Figure 2.1. The twisting is done to reduce the amount of interference one wire can inflict on the other. Recall from physics two important laws: (1) a current passing through a wire creates a magnetic field around that wire; and (2) a magnetic field passing over a wire induces a current in that wire. Therefore, a current in one wire can produce an unwanted current in a second wire, known as crosstalk, a condition that is minimized or eliminated by twisting the wire. In addition to, or in place of, physically twisting the wires, a protective shielding can be added to one or more of them. Typically a metal braid, this shielding provides a level of isolation from stray magnetic fields. Therefore, twisted pair cables can be purchased shielded or unshielded. For comparison purposes, unshielded twisted pair cable can transfer data at 10 megabits per second (Mbps) for short distances, such as 100 meters, and then only through benign environments. At a slower data rate of 9600 bps, a distance of 7 to 8 kilometers is possible.

Note that the distance a signal can travel is inversely proportional to the speed of transmission and level of noise in the environment. This property is explained more fully in the section on errors later in the chapter. A second distinction is that the smaller the diameter of the wire (its gauge), the greater is its resistance to the transfer of a signal. Greater resistance produces a lower transmission rate. A smaller gauge wire offers less total surface for the signal, producing an increased signal loss. For example, 24-gauge wire (standard telephone twisted wire) under laboratory conditions can transfer a 9600 bits per second (bps) signal for approximately 12 kilometers, while a 19-gauge wire (larger diameter) can transfer 9600 bps for 22.5 kilometers. Real-life environments can produce ranges 30 percent to 50 percent less than laboratory conditions.

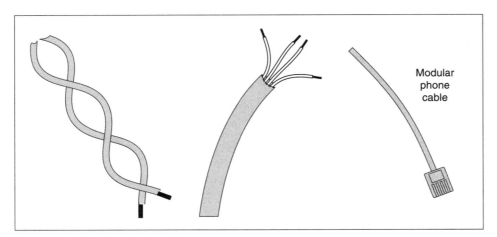

FIGURE 2.1   Examples of twisted pair wires

Shielded twisted pair cable can carry data signals at higher data rates and for longer distances than the unshielded type since the shielding provides a level of protection from extraneous magnetic interference. Unshielded twisted pair cable is initially less expensive than the shielded type, but it may be more expensive in the long run because of its limited transmission speed and distance.

A second distinction in twisted pair cable is whether the wire is privately or publicly owned. Typically, when the twisted pair remains within the confines of one room or one building, the wire is privately owned. Occasionally when the twisted pair runs between buildings and especially if it must cross public property, private cable is not used. In this case public twisted pair cable from the local telephone company is used.

There are basically two types of public twisted pair cable: switched (or dial up) and leased. Switched public twisted pair cable is what connects most homes and businesses. This is the phone line that crosses the city, enters your building, and connects to your telephone. It can essentially go anywhere the telephone lines go at speeds up to 19,200 bps. Leased public twisted pair cable is a dedicated phone line provided by the telephone company under contract provisions. This dedicated phone line is established between sender and receiver and is "conditioned" (electronic equipment is added to the line to reduce noise) so that it is quieter, thus introducing fewer errors into the line. Leased public twisted pair cable can run from anywhere to almost anywhere (depending on the telephone company) and is capable of carrying data up to 64,000 bps.

Coaxial Cable

Coaxial cable, in its simplest form, is a single wire wrapped in a foam insulation, surrounded by a braided metal shield, then covered in a plastic jacket (Figure 2.2). Because of its good shielding properties, coaxial cable is capable of carrying signals at much higher frequencies than twisted pair.

Coaxial cable can be divided into two major technologies depending on the type of signal it carries: baseband or broadband. **Baseband coaxial** technology uses digital signaling in which the cable carries only one channel of data. **Broadband coaxial** technology employs an analog signal and is capable of supporting multiple channels of data simultaneously. Baseband coaxial cable can typically carry signals of 10 Mbps and

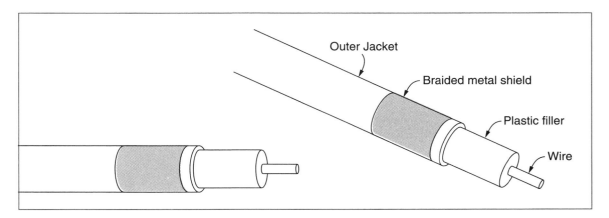

FIGURE 2.2   Coaxial cable wire

needs repeaters at least every kilometer. A repeater is an electronic device that regenerates a weakened digital signal into a clean, strong digital signal. Broadband coaxial cable can also carry signals of 10 Mbps and requires amplifiers at least every kilometer. An amplifier is an electronic device that regenerates a weakened analog signal into a clean, strong one. Since broadband can carry multiple channels at one time, its overall capacity is much greater than baseband. Typical coaxial cable systems can carry as many as 10,800 voice-grade channels. Consider the coaxial cable that transmits cable television into homes. Many cable companies offer 30 or more channels, with each channel carrying the equivalent of millions of bits per second of information.

Fiber Optic Cable

All the conducted media discussed above have one great weakness: electromagnetic interference. Electromagnetic interference is the electronic distortion a metal wire experiences when a stray magnetic field passes over it. This interference can be reduced with shielding, but it cannot be completely eliminated unless you use fiber optic cable. **Fiber optic cable** is a thin glass cable approximately the size of a human hair surrounded by a plastic coating to provide flexibility. A light source is placed at the transmitting end and quickly switched on and off. The light signal travels down the glass cable and is detected by an optic sensor on the receiving end. The light source can be either a simple light-emitting diode (LED) or a more complex laser. While the LED is less expensive, a laser can produce much higher data transmission rates. Fiber optic cable is quite capable of transmitting data at over 2 Gbps (2 gigabits or 2 *billion* bits per second) with a need for a repeater only every 32 to 48 kilo-meters (km).

Since fiber optic cable passes electrically nonconducting photons through a glass medium, it is immune to electromagnetic interference and quite difficult to wiretap. The medium's very high data transfer rates and extremely low error susceptibility make believable Bell Laboratories' claim of successfully placing 30,000 simultaneous phone calls over a single fiber optic cable!

One disadvantage of fiber optic cable is the difficulty of accurately joining together two separate pieces of cable. A second disadvantage is the relatively high costs of the cable and supporting electronic equipment. A summary of conducted media with relative costs, distances, and data transmission speeds is shown in Table 2.1.

Radiated Media

Radio, visible light, X-rays, and gamma rays are all examples of electromagnetic waves or electromagnetic radiation. Electromagnetic radiation is energy propagated through space or material media in the form of an advancing disturbance in electric and magnetic fields existing in space or in the media. Radio waves, generally speaking, are emitted by the accelerations of free electrons, as in a radio antenna wire. The basic difference between these various types of electromagnetic waves is their differing wavelengths, or frequencies (Figure 2.3). Concerning the transfer of data, we examine five basic groups: (1) broadcast radio (AM and FM), (2) packet radio, (3) cellular phone systems, (4) terrestrial microwave, and (5) satellite microwave.

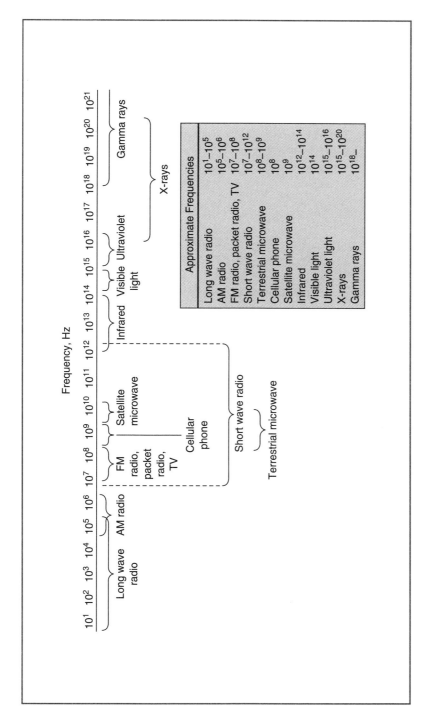

FIGURE 2.3   Electromagnetic wave frequencies

TABLE 2.1   Summary of the Conducted Media Types

Cable	Initial Costs[1]	Distances Before Regeneration	Transmission Speeds
Unshielded twisted Pair—private	$.16–1.00/ft	100 m 7–8 km	10 Mbps 9600 bps[2]
Shielded twisted Pair—private	$.16–1.00/ft	200 m 10–15 km	10 Mbps 9600 bps[2]
Twisted pair public	[3]	Handled by phone company	2400 bps– 1.544 Mbps[2]
Coaxial baseband	$.65–several $/ft	1 km	10 Mbps[2]
Coaxial broadband	$.65–several $/ft	1 km	10 Mbps[2]
Fiber optic	$1.00–several $/ft	32–48 km	>2 Gbps

[1]Approximate values, given for comparison purposes.
[2]Highly dependent on environmental factors.
[3]Initial costs handled by phone company. The phone company bills the user according to contract provisions.

Broadcast Radio

Broadcast radio is an omnidirectional transmission of radio frequency waves from a single point. Used primarily for the transmission of music and voice, broadcast radio has also been used for specialized pagers, the Alohanet (a data transfer system used in Hawaii), and Videotex (information transmission using video). The current discussion centers on the use of FM radio waves to transmit or sell microcomputer software to subscribers.

Since the ionosphere reflects back to earth radio signals of less than 30 megahertz (MHz, or 1 million cycles per second), AM radio (300 kilohertz [KHz or 1000 cycles per second] to 3000 KHz) is capable of transmitting over distances much greater than line of sight (which is approximately 80 kilometers, given a 100-meter-high transmission tower). FM radio, 30 MHz to 300 MHz, is capable only of line-of-sight transmissions. AM radio transmissions typically suffer from amplitude distortion, such as that caused by electrical storms and high-power lines. FM radio transmissions suffer from multipath interference, in which one signal is split into two by a building or a mountain, and then the pieces rejoin out of phase. Data transmission rates using AM and FM radio signals are typically no greater than 9600 bps.

Packet Radio

Packet radio is essentially the same as FM radio. Data is broken into fixed-sized packets and transmitted at the same frequencies, bps, and distances as FM radio.

Cellular Phone Systems

Cellular phone systems are used primarily to transmit voices and, to a smaller degree, to send data to and from mobile vehicles. The range of a cellular system is limited to the geographic area served by the cellular phone company.

Terrestrial Microwave

Terrestrial microwave systems are used primarily for long-distance telephone and video transmissions. These use line-of-sight transmission in the 300 MHz to single-digit-GHz (gigahertz, or 1 billion hertz) frequency range. Data transmission speeds often reach 45 Mbps. The basic disadvantages of terrestrial microwave are a loss of signal or attenuation, and interference from other signals.

Satellite Microwave

Satellite microwave is essentially terrestrial microwave except that the signal travels from earth to a satellite and back to earth, thus achieving much greater distances than line-of-sight transmission. A satellite in geosynchronous orbit (meaning that it remains fixed over one location on the earth) can transmit signals approximately one-third the distance around the earth. Frequencies are typically in the 3 GHz to 30 GHz range with a data transfer rate of 100 Mbps. The same disadvantages that occur with terrestrial microwave apply to satellite microwave. A summary of radiated media is shown in Table 2.2.

MEDIA SELECTION CRITERIA

In preparation for data transmission, the selection of one type of medium over another is an important issue. Many projects have performed miserably and even failed because of a decision to use an inappropriate type of medium. Assuming you have the option of choosing a medium, you should consider many factors before making that final choice. The sections below present the principal factors you should consider in your decision.

Cost

There are costs associated with all types of media. While the values cited in this text are approximate, they can help you compare the relative costs of one medium with

TABLE 2.2   Summary of Radiated Media Types

Radiated Media Types	Frequencies	Distances	Transmission Rate
AM transmission	300 kHz–3,000 kHz	>200 km	9600 bps
FM transmission	30 mHz–300 mHz	80 km	9600 bps
Packet radio	30 mHz–300 mHz	80 km	9600 bps
Cellular phone	800 mHz–900 mHz	geographic market	9600 bps
Terrestrial microwave	300 mHz–9 gHz	80 km	45 Mbps
Satellite microwave	3 gHz–30 gHz	1/3 circumference of earth	>100 Mbps

another. For example, twisted pair cable is generally less expensive than coaxial cable, which is generally less expensive than fiber-optic cable. There are also different types of cost. There may be initial costs, transmission costs, and maintenance costs for each type of medium. While it is quite easy to browse catalogs for initial costs of twisted pair and coaxial cable, it is more difficult to determine the maintenance costs of a particular medium two, five, or ten years down the road. Unfortunately, initial cost is too often the main limiting factor to considering better types of media.

Speed

To evaluate media properly you should consider two types of speed: data transmission speed and propagation speed. Data transmission speed is typically calculated as the number of bits per second that can be transmitted. Maximum bits per second for a particular medium depends proportionally on the effective bandwidth (in cycles per second, or Hertz) of that medium, and the distance over which the data must travel. If you need an implementation with a minimally acceptable bits per second speed, the chosen medium would have to support a proportionate frequency for the necessary distance.

Propagation speed is the speed at which a signal moves through a medium. For electrically conducted media, this speed is very near the speed of light (2×10^8 meters/second). For fiber optic cable and radiated media, propagation speed is the speed of light (2.998×10^8 m/s). While this speed seems "fast enough" for most applications, keep in mind that the propagation speed to send a signal to a geosynchronous satellite and back to earth is approximately 0.5 seconds. If you are sending data overseas, this 0.5 second delay is more than likely acceptable. If you are transferring data from one state to another, you might want to consider a conducted medium (if possible) rather than the 0.5-second satellite delay.

Expandability

Certain media lend themselves more easily to expansion. For example, it is relatively simple physically to tap into coaxial cable but quite difficult to splice into fiber optic cable. In a different case, public twisted pair cable can go anywhere the phone company can go, but private twisted pair cable cannot cross public land.

Distance

The conditions of distance are similar to those of expansion in that various media are physically capable of carrying a signal further with less introduction of noise or errors. Fiber optic cable is the king of long distance conducted media as it can run for great lengths with very little noise introduced.

Environment

Another factor that must be considered in the selection process is the environment. Many types of environments are hazardous to certain media. Industrial environments with heavy machinery produce electromagnetic radiation that can interfere with improperly shielded cables. For radiated media, electrical storms are very disruptive to AM signals, while tall buildings and mountains can distort FM signals. Even sunspots can disrupt radio waves.

Maintenance

All types of media require some degree of maintenance. For conducted media the most common maintenance required is repair of broken or faulty lines. If a line breaks and needs to be spliced, twisted pair and coaxial cable can be spliced more easily than fiber optic. Various connectors used on the ends of the different conducted media are often prone to failure. Public twisted pair cable, operated and owned by a telephone company, can be affected by inclement weather, downed telephone poles, and switching equipment failure. Many more examples can be listed, but the point is that the system designer should investigate maintenance issues thoroughly before committing to one type of medium.

Security

If data must be secured in transmission, it is important that the media not be easily tapped. All conductive media except fiber optic cable can easily be wire-tapped by someone listening to the electromagnetic signal traveling through the wire. Radiated media can also be listened to since the signal is simply a radio wave sent through the air. Secure data should be encrypted; then, even if someone taps into the line, he or she cannot understand the data.

DATA

Now that we have examined the media in some detail, we should discuss just what it is the media carry: information. The word *information* is too general a term for our use here. To be more precise, information that is transferred over a communications medium can be broken into two components: *data* and *signals.*

A usable definition of data is entities that convey meaning within a computer or computer system. The entities could be as simple as the digital 1s and 0s used for computer information, or as complex as an analog voltage waveform used to convey the meaning of human speech over a phone line. Data, therefore, can be divided into two distinct categories: analog data and digital data.

Analog data is represented by a continuous waveform (Figure 2.4). This waveform can take on an infinite number of values between its upper and lower bounds. The human voice, music, and noise are common examples of analog data. **Digital data** is represented by a discrete, or noncontinuous waveform, as shown in Figure 2.5.

SIGNALS

Signals can be defined as the electric or electromagnetic encoding of data. Like data, signals can be analog or digital. Typically, digital signals are used to convey digital

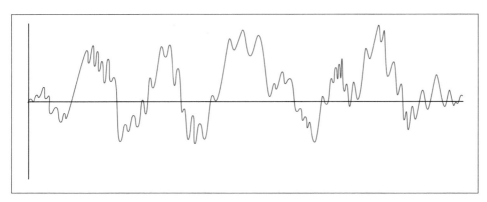

FIGURE 2.4   Analog data represented as a continuous waveform

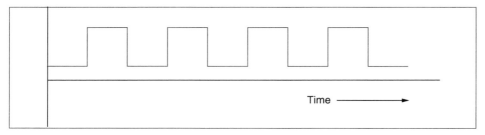

FIGURE 2.5   Digital data represented as a noncontinuous waveform

data, and analog signals are used to convey analog data. However, you can also use analog signals to convey digital data and digital signals to convey analog data. Because of the electronic equipment used to transmit a signal along a medium, the medium very often dictates the type of signals it can transmit. Certain media are capable of supporting only analog transmissions, while others can support only digital transmissions.

All signals, both analog and digital, consist of three basic parts: amplitude, frequency, and phase. Every signal can be written mathematically as a form of a sine wave:

$$s(t) = A \sin (2\pi f t + \theta)$$

where

$s(t)$ is the signal at time t

A is the amplitude

f is the frequency

θ is the phase

The **amplitude** of a signal is the height of the sine wave above (or below) a given reference point (Figure 2.6).

The **frequency** of a signal is the number of times a signal makes a complete cycle within a given time frame. The length, or time interval, of one cycle is called its **period**.

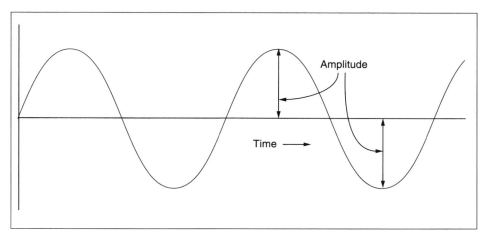

FIGURE 2.6   The amplitude of an analog signal

Figure 2.7c shows a sine wave completing three cycles in the given time period. If the time *t* is one second, the signal has completed three cycles in one second. Cycles per second is referred to as **Hertz (Hz)**. Thus, the signal in Figure 2.7c has a frequency of 3 Hz.

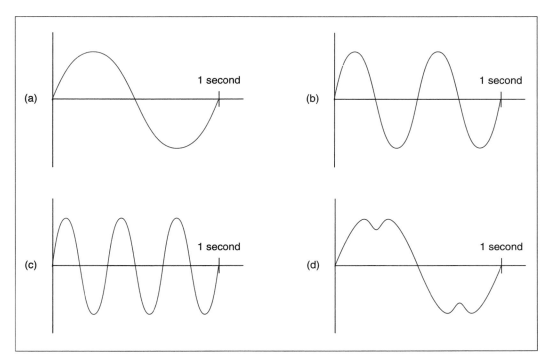

FIGURE 2.7   Three signals of 1 Hz, 2 Hz, and 3 Hz, and their composite signal

Many times a signal will take on a range of frequencies, from a given minimum to a given maximum, or one signal may be a **composite** of multiple signals, each with a different frequency. This range of frequencies is called its **spectrum**. If three signals of 1 Hz, 2 Hz, and 3 Hz are combined into one composite signal, the resulting sine wave would look like Figure 2.7d. The spectrum of Figure 2.7d would be 1 Hz to 3 Hz.

The **absolute bandwidth** of a signal is the absolute value of the difference between the lowest and highest frequencies. The absolute bandwidth of Figure 2.7d is 2 Hz. The **effective bandwidth** is a more commonly found and realistic term. It is typically less than the absolute bandwidth because of extraneous noise that degrades the original signal. Many professionals rely more on the effective bandwidth than the absolute bandwidth when making communication decisions since most situations deal with the real-world problems of noise and interference.

Perhaps one of the most important concepts of this chapter is the following: **the greater the frequency of a signal, the higher the possible data transfer rate; the higher the desired data transfer rate, the greater the needed signal frequency.** We can see a direct relation between the frequency of a signal and the data rate (in bits per second, or bps) of the data that a signal can carry. As an example, examine the oversimplified signal in Figure 2.8a. Within the time period of one second, the signal has a frequency of 2 Hz. If we consider the wave above the *X* axis as representing one binary value, and the wave below the axis as representing another binary value, this signal

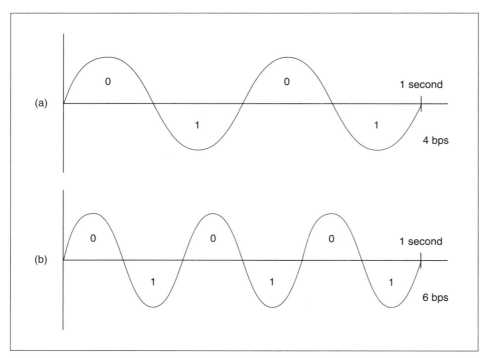

FIGURE 2.8   Simplified graphic representation of relationship of frequency to bits per second (bps)

transmits four binary values, or bits, in one second. If we increase the frequency of the signal as we have done in Figure 2.8b, we can now transmit more binary values in the same amount of time. By increasing the frequency of the signal, we have proportionally increased the data transfer rate.

This is an important concept. Later we will see particular applications demanding a minimum data transfer rate. To carry this minimum data transfer rate, a signal must have a proportionally equivalent frequency. Every medium has a recommended range of frequencies in which it performs best. To transfer data at a high rate of speed, you will need a signal with an equally high frequency, so you will need to select a medium designed for carrying such a frequency. As always, you don't get something for nothing. Media capable of supporting higher frequencies generally cost more than those that do not support them.

A simple but elegant calculation for determining the data transfer rate of a signal given its frequency is **Nyquist's theorem**:

$$C = 2f \log_2 L$$

where C is the channel capacity in bits per second, f is the frequency of the signal, and L is the number of signaling levels. Given a 3100 Hz signal and two signaling levels (one level indicates a binary 1, the second indicates a binary 0), the resulting channel capacity is 6200 bits per second. While this is a simple calculation, one major factor has been left out of the equation. Noise always reduces the effective data transfer rate of a signal. **Shannon's formula** calculates the data transfer rate of a given signal and incorporates noise into the equation:

$$S(f) = f \log_2 (1 + W/N) \text{ bps}$$

where

$S(f)$ is the data transfer rate in bits per second

f is the frequency of the signal

W is the power of the signal in watts

N is the power of the noise in watts

As an example, consider a 3100 Hz signal with a power level of 500 watts and a noise level of 100 watts:

$$S(f) = 3100 \times \log_2 (1 + 500/100) \text{ bps}$$
$$= 3100 \times \log_2 (6) \text{ bps}$$
$$= 3100 \times 2.585 \text{ bps}$$
$$= 8014.2 \text{ bps}$$

Compare the results of this calculation using Shannon's formula with the previous calculation using Nyquist's theorem. If noise decreases the effective data transfer rate, why does Shannon's formula produce a higher transmission rate? The best answer we can provide at this time is that both formulas are approximations, and many other factors determine the rate of data transfer given a signal's frequency.

Signal Strength

A signal, while traveling through a medium, always experiences some loss of signal strength, or **attenuation**. Attenuation in a conducted medium is typically a logarithmic loss and is a function of distance. (Attenuation in a radiated medium is more complex and is not discussed here.) To measure the logarithmic loss of a signal, we use the term **decibel (dB)**. The decibel is a **relative** measure of loss (or gain), and is expressed as

$$N_{dB} = 10 \log_{10} (P_2/P_1) \text{ dB}$$

where P_2 and P_1 are the ending and beginning power levels of the signal. For example, if a signal starts at a transmitter with 10 watts of power and arrives at a receiver with 5 watts of power, the signal loss in dB is calculated as follows:

$$N_{dB} = 10 \log_{10} 5/10 \text{ dB}$$
$$= 10 \, (-0.3) \text{ dB}$$
$$= -3 \text{ dB}$$

Consequently, there is a 3 dB loss (or −3 dB gain) between sender and receiver. Remember that the decibel is a relative measure of loss or gain. You cannot take a single power level at time t and compute the decibel value of that signal without having a reference or beginning power level.

The **decibel watt (dBW)** is an absolute measure of a signal and uses a reference level of 1 watt, as shown for dBW in the following equation:

$$dBW = 10 \log_{10} (W/1)$$

where W is the power level in watts.

Since attenuation is a logarithmic loss and the decibel is a logarithmic value, to calculate the overall loss or gain of a system is as easy as adding all the individual decibel losses and gains. For example, in Figure 2.9 we see a communication line running from point A, through point B, and ending at point C. We know that the communication line

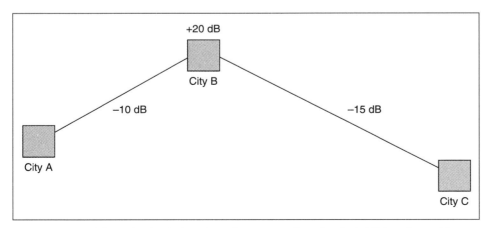

FIGURE 2.9   Hypothetical situation demonstrating decibel (dB) gain and loss

from A to B experiences a 10 dB loss, point B has a 20 dB amplifier, and the communication line from B to C experiences a 15 dB loss. What is the overall gain or loss of the signal? To calculate, simply add all dB gains and losses: −10 dB (A to B link) + 20 dB (amplifier) + −15 dB (B to C link) = −5 dB.

One final term that is fairly important is the **signal to noise ratio (S/N_{dB})**. As its name implies, it is the ratio of signal power to noise power for any given signal.

$S/N_{dB} = 10 \log_{10}$ (signal power/noise power) dB

If we know that the power level for a particular signal is 1000 watts and the accompanying noise level is 20 milliwatts, the signal to noise ratio is this:

$$S/N_{dB} = 10 \log_{10} (1000 / 0.020) \text{ dB}$$
$$= 10 \log_{10} (50{,}000) \text{ dB}$$
$$= 10 \times 4.69 \text{ dB}$$
$$= 46.9 \text{ dB}$$

Digital Data and Digital Signals

We have previously seen that data may be digital or analog. Similarly, the signals that convey the data may be digital or analog. You might think that digital signals are always used to convey digital data, and analog signals are always used to convey analog data, but this is not always so. The choice of using either analog or digital signals often depends on the medium that is used and the environment in which the signals must travel. For example, the telephone system from inception has been an analog environment. While twisted pair wire is capable of carrying either analog or digital signals, the electronic equipment used to amplify and equalize the lines can accept only analog signals. Therefore, it is not uncommon to transmit digital data over telephone lines using analog signals. The opposite situation—transmitting analog data with digital signals—is also fairly common. These conditions yield four combinations of data and signals: digital data transmitted using digital signals, digital data transmitted using analog signals, analog data transmitted using analog signals, and analog data transmitted using digital signals. Let's first consider transmitting digital data using digital signals. To do this, the logical 1s and 0s of the digital data must be converted to a physical form that can be transmitted over the medium. If the medium is twisted pair cable, for example, the logical 1s and 0s could be converted to the absence or presence of voltage on the medium. Thus, if we wish to transmit a logic 1, we could transmit no voltage on the medium. If we wish to transmit a logic 0, we could transmit a positive voltage. While this is a very simple example, most digital encoding techniques work in a similar fashion. As a first example, consider the NRZ-L (**N**on-**R**eturn to **Z**ero-**L**evel) digital encoding format shown in Figure 2.10.

A logic 0 in the data is represented as a "high" voltage signal, while a logic 1 in the data is represented as a "low" voltage signal. In this example, the digital signal simply appears as the logical opposite of the digital data.

This is not the case with the NRZ-I (**N**on-**R**eturn to **Z**ero, **I**nvert on ones) or RZ (**R**eturn to **Z**ero) digital encoding formats. The NRZ-I format has a voltage change at the beginning of a 1 and no voltage change at the beginning of a 0. The RZ scheme

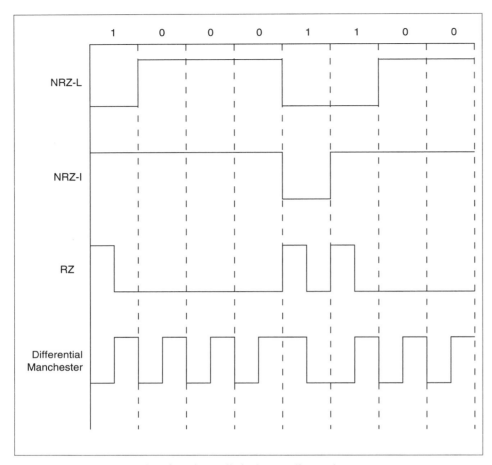

FIGURE 2.10   Example of various digital encoding schemes

employs a low to high signal transition at the beginning of a logic 1 followed by a high to low signal transition in the middle of the bit. A logic 0 has no signal transitions.

An inherent problem with the first three schemes is that long sequences of 0s produce a nonchanging linear signal. Often the receiver "looks for" signal changes so that it may synchronize its internal clock with the incoming signal. If the incoming signal consists of a long stream of 0s, how can the receiver synchronize itself? The **Manchester** class of digital codes ensures that each bit has some type of signal change. In this way, the receiver expects a signal change at regular intervals and synchronizes itself with the incoming stream. The **Differential Manchester** code shown in Figure 2.10 exhibits a signal change in the middle of each bit. To distinguish 1s from 0s, each 0 bit starts with a signal transition; 1s do not have a signal transition at the beginning of the bit.

While the Manchester codes provide good synchronization capabilities, they have one serious drawback: the modulation rate is twice the bit rate. Stated another way, the baud rate is twice the bps. Many people erroneously believe that baud and bps (bits per second) are two terms for the same thing. The **baud rate** of a signal is the number of

signal changes per second, while bps is simply the number of bits per second. As shown in the Differential Manchester example, the signal changes twice during a logic 0, once at the beginning of the bit, and once in the middle of the bit. If you wish to transmit 100 million logic 0s per second using Differential Manchester encoding, the signal will have to change 200 million times per second. The baud rate is twice the bits per second. Likewise, the hardware must be capable of reacting to not 100 million but *200 million* transitions per second. As always, you don't get something for nothing. The hardware that can handle this transmission is quite elaborate and expensive.

An encoding scheme that tries to satisfy the synchronization problem while avoiding the "baud equals two times the bps" problem is the newer **4B/5B code** commonly used for fiber-optic systems. The first step in generating the 4B/5B code is to convert four-bit quantities of the original data into new five-bit quantities. Five bits yield 32 combinations, of which only 16 are used. The 16 combinations were chosen so that no code has three or more consecutive 0s. If we then transmit the five-bit quantities using NRZ-I encoding, we will never transmit more than two 0s in a row. (Recall that a 0 in NRZ-I means no voltage change at the beginning of the 0 bit.) Figure 2.11 shows the 4B/5B code in detail. By converting a four-bit code to a five-bit code, we have introduced a 20 percent overhead, which is much better than Differential Manchester's 100 percent overhead.

Digital Data and Analog Signals

Since the U.S. telephone system is a predominantly analog medium, the transfer of digital data over a phone line requires conversion of the data to an analog signal. The

Valid Data Symbols		Invalid codes
Original 4-bit data	New 5-bit code	
0000	11110	
0001	01001	
0010	10100	
0011	10101	
0100	01010	00001
0101	01011	00010
0110	01110	00011
0111	01111	01000
1000	10010	10000
1001	10011	
1010	10110	
1011	10111	
1100	11010	
1101	11011	
1110	11100	
1111	11101	

0110 → Becomes → 01110 → Transmitted as → 0 1 1 1 0

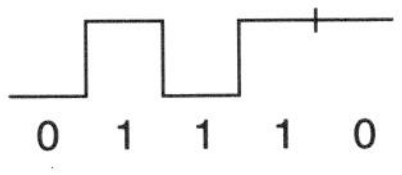

FIGURE 2.11   The 4B/5B digital encoding scheme

technique of converting digital data to an analog signal is **modulation**. A device that modulates digital data onto an analog signal and then demodulates the analog signal back to digital data is a **modem** (**MO**dulator/ **DEM**odulator). The modem and its characteristics are discussed in detail in Chapter 3, but here we examine the modulation techniques that are currently popular.

The simplest modulation technique is **amplitude modulation**. As shown in Figure 2.12a, logic 1 and logic 0 are represented by two different amplitudes of a signal. The higher amplitude represents one bit value; the lower amplitude represents the second bit value. Amplitude modulation has a weakness in that it is susceptible to sudden noise impulses such as the static charges created by a lightning storm. It is also considered one of the least efficient modulation techniques and typically does not exceed 1200 bps over voice-grade lines.

Frequency modulation uses two different frequency ranges to represent a logic 1 and logic 0 (Figure 2.12b). For example, the higher frequency signal might represent a logic 1, while the lower frequency signal might represent a logic 0. A **full duplex** connection (both ends of the communication line transmitting data to each other simultaneously) uses two sets of these frequency ranges. The standard Bell System 103F series modem uses 1700 Hz as the cutoff frequency between the two directions (Figure 2.13), with one direction using 1070 Hz as a logic 0 and 1270 Hz as a logic 1, and the second direction

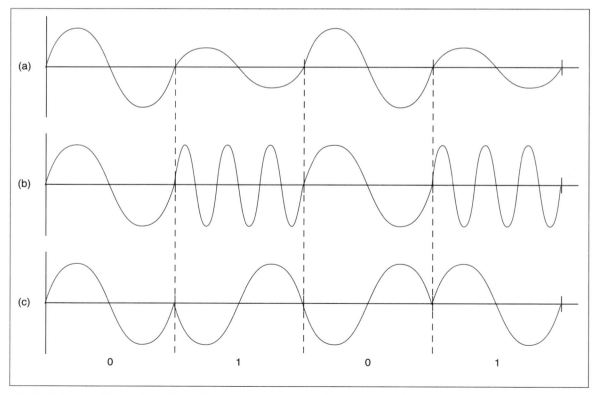

FIGURE 2.12   Examples of amplitude, frequency, and phase modulation

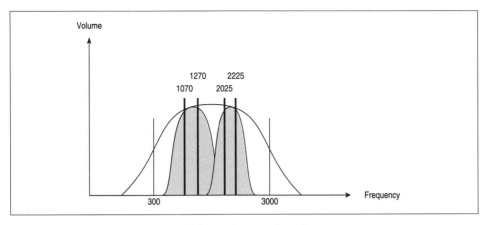

FIGURE 2.13   Bell Systems 103F series modem frequency ranges

using 2025 Hz as a logic 0 and 2225 Hz as a logic 1. Frequency modulation is less susceptible to noise than is amplitude modulation (witness the comparative quality of FM radio over AM radio) and can be used at fairly high frequencies (3 MHz to 30 MHz).

The third modulation technique, **phase modulation**, represents logic 0 and logic 1 by a shift in the phase of the signal. In simple phase modulation, a logic 0 is no phase shift, while a logic 1 is a phase shift of 180 degrees. This type of phase shift is termed **differential phase shift** since the 0 or 1 depends on whether there was a phase shift between the previous bit transmitted and the current bit transmitted (see Figure 2.12c).

Phase modulation is susceptible to fewer errors than either amplitude or frequency modulation and thus can be used at higher frequencies. Phase modulation is so accurate that the user can increase efficiency by introducing multiple phase shift angles. **Quadrature phase modulation** goes further than simple phase modulation by incorporating four different phase angles:

45-degree phase shift represents a logic 11

135-degree phase shift represents a logic 10

225-degree phase shift represents a logic 01

315-degree phase shift represents a logic 00

In this case, each phase shift represents two bits, doubling the efficiency of simple phase modulation. With this technique, we see our first example of one signal change equaling two bits of information, that is, 1 baud equals 2 bps.

This technique can be carried even further by using 12 different phase shift angles with a combination of two different amplitudes. Figure 2.14 shows only four of the phase angles using the two different amplitudes, resulting in 16 combinations. This technique, **amplitude phase modulation**, has four bits representing each signal change. Therefore, the bps of the data transmitted using this technique is four times the baud rate. By using a signal that changes 2400 times per second (baud rate of 2400), we have achieved a data transfer rate of 9600 bps. This technique is commonly employed in modern-day modems (Chapter 3).

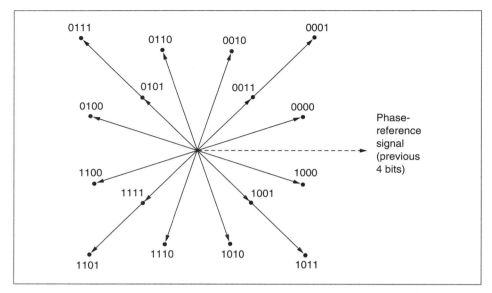

FIGURE 2.14   Graphic representation of amplitude phase modulation

The last modulation technique we examine here is **Trellis coded modulation**, a form of forward-error correction. Without getting into the detailed mathematics, let us simply say here that a fifth redundant bit (for error correction) is added to a 4-bit code. Each signal change is represented by 5 bits, but a baud rate of 2400 still yields a data transfer rate of only 9600 bps. The redundant bit provides an extra level of error protection, but at a higher cost.

Analog Data and Digital Signals

Many times it is necessary to transmit analog data over a digital medium. For example, many scientific laboratories contain testing equipment that generates test results as analog data and then transmits them to a computer that collects and stores the data. One technique that converts analog data to a digital signal is **pulse code modulation**. The hardware that converts the analog data to a digital signal follows the analog waveform, taking "snapshots" of the analog data at fixed intervals. A snapshot involves calculating the height (determining the voltage) of the analog waveform above a given threshold. The height, a decimal value, is converted to an equivalent fixed-size binary value. This binary value is then transmitted by means of a digital encoding format.

Figure 2.15 shows that at time t (X axis), a snapshot of the analog waveform is taken, resulting in the decimal value 12 (Y axis). The 12 is then converted to a 16-bit binary value (0000 0000 0000 1100) and transmitted to a device to be stored. (Sixteen-bit conversion is quite common in the industry.) At time $2t$, a second snapshot is taken and the decimal value 4 is converted to a 16-bit binary value and stored. This process continues as snapshots are taken, converted to binary form, and stored. If you want to reconstruct the analog waveform from the stored digital values, special hardware

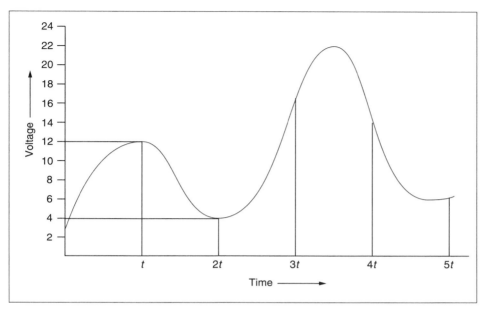

FIGURE 2.15   Example of taking "snapshots" of an analog signal for conversion to a digital signal

converts each 16-bit binary value back to decimal and generates an electric pulse of appropriate magnitude (height). With a continuous incoming stream of converted values, something close to the original waveform should result (Figure 2.16). Note that the closer the snapshots are taken to one another (the smaller the time intervals, or the finer the resolution), the more accurate will be the reconstructed waveform (Figure 2.17).

As always, you don't get something for nothing. To take the snapshots at shorter time intervals, the hardware must be of high enough quality to track the incoming signal quickly and perform the necessary conversions.

A second noteworthy item: the greater the number of bits used in the binary value, the larger will be the range of decimal integer values or the greater will be the precision of decimal fractional values. This increase in precision also requires more elaborate hardware and a higher number of bits transmitted for each sampled signal.

A second example of analog data to digital signal conversion is **delta modulation**. With delta modulation, the electronic hardware "tracks" the analog signal. During each time period t, the hardware determines whether the waveform has risen one delta step or dropped one delta step. If the waveform rises, a digital 1 is transmitted. If the waveform drops, a digital 0 is transmitted. With this technique, fewer bits are generated, since the hardware produces only a 1 or 0 for each time interval t. Unfortunately, if the analog waveform rises or drops too quickly, the hardware may not be able to keep up with the change, and **slope overload noise** will result (assuming t and δ are held constant). Another problem occurs when the analog waveform does not change at all. Since the hardware outputs 1 and 0 only for rise and fall, a nonchanging waveform generates alternating 1s and 0s, called **quantizing noise** (Figure 2.18).

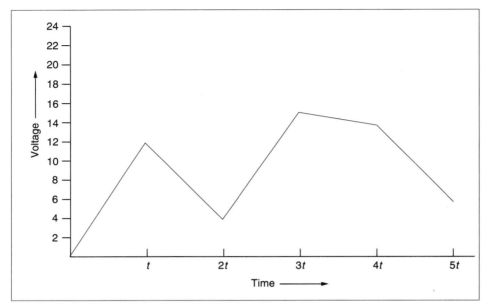

FIGURE 2.16   Reconstruction of analog waveform from digital "snapshots"

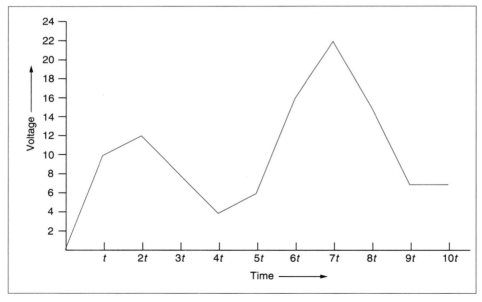

FIGURE 2.17   Reconstruction of analog waveform using more digital samples

Analog Data and Analog Signals

The final grouping of analog data and analog signals is similar to the combination of digital data and digital signals: if analog data is transmitted using analog signals, very

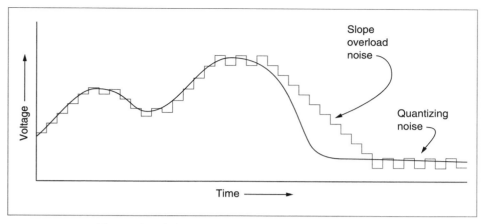

FIGURE 2.18   Example of slope overload noise and quantizing noise

little needs to be done to the original analog data. The most common change that takes place with analog data is for it to be modulated into a different frequency analog signal. In this way, the original analog data can be moved to a different frequency range so that it may fit on an appropriate medium, or share a medium with other analog signals at other frequencies (multiplexing).

DATA CODES

One of the most common forms of data transmitted between a sender and receiver is textual data. For example, if one banking institution wishes to transfer money to a second banking institution, much information is often transmitted between the two institutions, such as account numbers, names of account owners, bank names, addresses, and the amount of money to be transferred. This information is transmitted as a sequence of characters. To recognize one character from another, each character has a unique pattern of 1s and 0s. The pattern selected is the **data code**. There are many data codes, but here we examine three of the more commonly used ones: Baudot, EBCDIC, and ASCII.

Baudot Code

The Baudot code was developed by Emile Baudot and uses 5-bit patterns to represent the characters A to Z, the numbers 0 to 9, and several special characters. A simple calculation demonstrates, however, that a 5-bit pattern yields only 32 combinations ($2^5 = 32$). With 26 letters of the alphabet and 10 digits we have 36 distinct characters, and we haven't even included the special characters. How can the Baudot code recognize more than 32 characters with only a 5-bit code? As we see in Table 2.3, each 5-bit pattern can actually represent two different characters. For example, the bit pattern 10000 can be either the letter T or the number 5. The letter T is a downshift character, while the number 5 is an upshift character. If the LETTERS (or DOWNSHIFT) character code (11111) was recently transmitted, then the pattern 10000 represents the letter T. If the FIGURES (or UPSHIFT) character code (11011) was more recently transmitted, then the pattern 10000 represents the number 5. Thus, every character except for a space is either an upshift character or a downshift character. If you wish to transmit a series of upshift characters, then those characters are preceded by the FIGURES character. If you want to then transmit one or more downshift characters, the LETTERS character is first transmitted.

This process works much like the keyboard on a typewriter or terminal. If the CAPS LOCK character has recently been pressed, all characters transmitted will be *UPSHIFT* characters. If the CAPS LOCK key is released, all characters transmitted will be *downshift* characters.

While the Baudot code is fairly dated and no longer used, it does demonstrate how a large set of possible values can be represented with a small number of bit combinations. We refer to this technique later when we discuss other techniques of data transmission.

EBCDIC

The EBCDIC code, or Extended Binary Coded Decimal Interchange Code, is an 8-bit code allowing 256 possible combinations. These 256 combinations include all uppercase and lowercase letters, the 10 digits, a large number of special symbols and

TABLE 2.3   The Baudot Character Set

	Letters	Figures		Letters	Figures
Binary	Shift	Shift	Binary	Shift	Shift
00000	blank	blank	10000	T	5
00001	E	3	10001	Z	+
00010	LF	LF	10010	L	)
00011	A	-	10011	W	2
00100	space	space	10100	H	reserved
00101	S	'	10101	Y	6
00110	I	8	10110	P	0
00111	U	7	10111	Q	1
01000	CR	CR	11000	O	9
01001	D	WRU	11001	B	?
01010	R	4	11010	G	reserved
01011	J	BELL	11011	FIGURES	FIGURES
01100	N	,	11100	M	.
01101	F	reserved	11101	X	/
01110	C	:	11110	V	=
01111	K	(	11111	LETTERS	LETTERS

punctuation marks, and a number of control characters. The control characters can be used to provide control between a processor and an input/output device, such as Line Feed (LF) and Carriage Return (CR). Certain control characters can also provide data transfer control between a source and destination when you are using BISYNC or a similar transmission protocol (Chapter 4). The EBCDIC characters are shown in table form in Figure 2.19.

ASCII

The last code set we examine is ASCII (American National Standard Code for Information Interchange). ASCII has become a government standard in the United States and is one of the most widely used data codes in the world. ASCII is a 7-bit data code that allows for 128 possible combinations, including upper and lowercase letters, the digits 0 to 9, special symbols, and control characters. Since the byte is a common unit and consists of eight bits, ASCII characters usually include an eighth bit. This eighth bit can be used to provide parity error detection (Chapter 4), can provide for 128 additional characters defined by the application, or simply be a binary 0. Since the ASCII standard defines only a 7-bit code, Table 2.4 shows the characters and their corresponding 7-bit values.

Bits	4	0	0	0	0	0	0	0	0	1	1	1	1	1	1	1	1
	3	0	0	0	0	1	1	1	1	0	0	0	0	1	1	1	1
	2	0	0	1	1	0	0	1	1	0	0	1	1	0	0	1	1
	1	0	1	0	1	0	1	0	1	0	1	0	1	0	1	0	1
8 7 6 5																	
0 0 0 0		NUL	SOH	STX	ETX	PF	HT	LC	DEL			SMM	VT	FF	CR	SO	SI
0 0 0 1		DLE	DC$_1$	DC$_2$	DC$_3$	RES	NL	BS	IL	CAN	EM	CC		IFS	IGS	IHS	IUS
0 0 1 0		DS	SOS	FS		BYP	LF	EOB	PRE			SM			ENQ	ACK	BEL
0 0 1 1				SYN		PN	RS	UC	EOT					DC$_4$	NAK		SUB
0 1 0 0		SP												<	(	+	|
0 1 0 1		&										!	$	.	)	;	¬
0 1 1 0		–												%	-	>	?
0 1 1 1														@	'	=	"
1 0 0 0			a	b	c	d	e	f	g	h	i						
1 0 0 1			j	k	l	m	n	o	p	q	r						
1 0 1 0				s	t	u	v	w	x	y	z						
1 0 1 1																	
1 1 0 0			A	B	C	D	E	F	G	H	I						
1 1 0 1			J	K	L	M	N	O	P	Q	R						
1 1 1 0				S	T	U	V	W	X	Y	Z						
1 1 1 1		0	1	2	3	4	5	6	7	8	9						

FIGURE 2.19   The EBCDIC character set

TABLE 2.4 **The ASCII Character Set**

		Bit positions 5, 6, 7							
		000	100	010	110	001	101	011	111
	0000	NUL	DLE	SP	0	@	P	'	p
	1000	SOH	DC1	!	1	A	Q	a	q
	0100	STX	DC2	"	2	B	R	b	r
	1100	ETX	DC3	#	3	C	S	c	s
	0010	EOT	DC4	$	4	D	T	d	t
	1010	ENQ	NAK	%	5	E	U	e	u
Bit	0110	ACK	SYN	&	6	F	V	f	v
positions	1110	BEL	ETB	'	7	G	W	g	w
1, 2, 3, 4	0001	BS	CAN	(	8	H	X	h	x
	1001	HT	EM	)	9	I	Y	i	y
	0101	LF	SUB	*	:	J	Z	j	z
	1101	VT	ESC	+	;	K	[	k	{
	0011	FF	FS	,	<	L	\	l	}
	1011	CR	CS	_	=	M	]	m	}
	0111	SO	RS	.	>	N	^	n	~
	1111	SI	US	/	?	O	-	o	DEL

ERRORS

As you might expect, many things can go wrong during the transmission of data. From a simple "blip" to a massive outage, transmitted data is susceptible to numerous types of errors. Therefore, you must take precautions to reduce the possibility of errors; when this is done, the receiver should be able to detect the error if the data is corrupted during transmission. In this section we examine the many types of errors that can occur in a data transmission stream. Chapter 4 addresses the problem of detecting and correcting errors.

Types of Errors

While the internetwork of phone lines, satellite systems, and radio networks is an engineering marvel, the system can still introduce errors. Conducted media have traditionally been plagued with many types of interference and noise. Satellite and radio networks are also prone to interference and crosstalk. Even the near-perfect fiber-optic cables can introduce errors into the system, though the probability of this happening is less than with the other types of transmission. In the rest of this section we examine several of the major types of errors that occur in transmission systems.

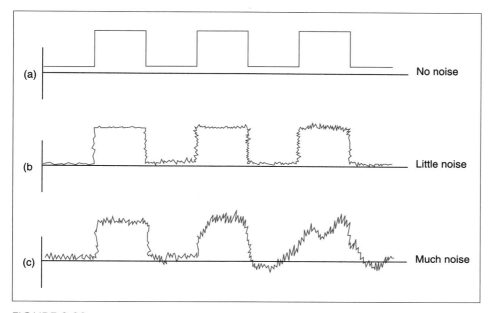

FIGURE 2.20   Effects of impulse noise on a digital signal

White noise (or thermal or Gaussian noise) is considered a relatively constant type of noise and is much like the static you hear when a radio is tuned between two stations. It is always present to some degree in transmission media and electronic devices and is temperature dependent. Since it is considered relatively constant, it can be reduced but never eliminated.

Impulse noise, or spikes, is a nonconstant noise and one of the most difficult errors to correct because it can occur randomly. The difficulty comes in separating the noise from the signal. Typically the noise is an analog burst of energy. If it interferes with an analog signal, removing it without destroying the signal is almost impossible. If impulse noise interferes with a digital signal, often the original digital signal can be recognized and recovered. Note in Figure 2.20b that the original digital signal can still be recognized despite the introduction of noise. This is not the case in Figure 2.20c where the amount of noise is so great that the original digital signal is not recognizable.

Crosstalk is an unwanted coupling between two different signal paths. This can be an electrical coupling as between two sets of twisted pair wire (as in a phone line) or as unwanted signals picked up by microwave antennas. Crosstalk is relatively constant and can be reduced with proper precautions and hardware.

Echo is the term applied to the reflective feedback of a transmitted signal as the signal moves through the medium. This error occurs most often at junctions where wires are connected or at the open end of a coaxial cable in a local area network. To minimize echo, a device called an echo suppressor can be attached to a line. An echo suppressor is essentially a filter that allows the signal to pass in one direction only. For local area networks, a small filter is placed on the open end of a wire to absorb any incoming signals.

Phase jitter and **timing jitter** are two similar types of errors. Phase jitter is a variation in the phase of a continuous signal resulting from the repeated phase changes introduced by phase modulation. Timing jitter is commonly found in ring local area networks (Chapter 6) when the clocked digital signal is repeated time after time as it passes through network stations. While both phase jitter and timing jitter can be reduced, they can almost never be removed completely. Electronic solutions for these two problems are beyond the scope of this text.

Delay distortion can occur because the velocity of propagation of a signal through a wire varies with the frequency of the signal. Since a signal may be composed of multiple frequencies, some of those frequencies may arrive at the destination sooner than others. Delay distortion can be particularly harmful to digital signals. With proper equalizing techniques, this type of error can be significantly reduced.

Attenuation is not necessarily a form of error but is the constant loss of a signal as it travels through a medium. As discussed in an earlier chapter, attenuation can be eliminated with the use of amplifiers for analog systems or repeaters for digital systems.

Error Prevention

With proper prevention techniques, many of the previous types of errors can be reduced. Nonetheless, you should not be lured into thinking that by simply applying various error prevention techniques, you will not encounter errors. You still need to implement an appropriate error detection method as introduced in Chapter 4. Without going into great detail, which should be left to the engineers and radio frequency specialists, we can examine some simple techniques for error prevention. The following may be applied to reduce the possibility of transmission errors.

1. Proper shielding of cables to reduce electromagnetic interference and crosstalk is essential.
2. Telephone line conditioning or equalization is provided by the phone company, and for an additional charge, it will provide various levels of conditioning to leased lines. This conditioning provides a quieter line more conducive to safe data transmission.
3. Replacing older equipment with more modern or digital equipment is initially expensive but often the most cost effective in the long run.
4. Proper use of digital repeaters and analog amplifiers can increase signal strength, thus decreasing the probability of errors and increasing the distance of transmission.
5. Reducing the number of devices, decreasing the length of cable runs, or reducing the transmission speed of the data may be effective ways to reduce the possibility of errors. While choices like these are not always desirable, sometimes they are the only choices available.

CASE STUDY _____

Chapter 1 introduced the hypothetical XYZ University and its planned computer center, which purchased a computer and 30 terminals. The mainframe computer, located in Building A (Figure 2.21), will be connected to six support personnel ter-

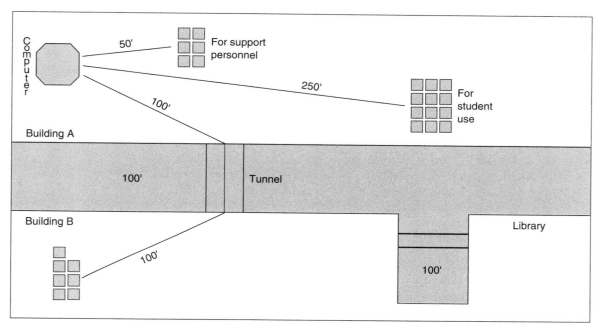

FIGURE 2.21   Layout of computer and 30 terminals for hypothetical computing center

minals located in Building A, twelve student-use terminals also located in Building A, and seven computer science faculty-use terminals located in Building B 100 feet away. There is a university-owned tunnel between buildings A and B. Both buildings contain laboratories and power systems that could introduce electromagnetic interference.

The first question to address is what type of media do we use to interconnect the terminals and mainframe computer? The first decision we can make is whether to use conducted media or radiated media. Given the relatively short distances involved, the fairly high costs of the radiated media such as terrestrial and satellite microwave, the fact that we can install private cables on university property, and the relatively low volume of data transferred between mainframe and terminal, we will rule out radiated media and concentrate on the conducted media.

Among the conducted media, our choices consist of twisted pair (shielded and unshielded) cable, coaxial cable, and fiber-optic cable. The issues to consider before choosing the media are these:

- *The lengths of the cable runs*—Will any cable run exceed the maximum recommended cable length? Given the current situation, the runs (longest is 300 feet) are short enough to fall safely within the limits of even the shortest-distance twisted pair cable.
- *The amount of electromagnetic interference*—Will any cable pass through an environment that introduces a high amount of electromagnetic interference,

thus jeopardizing the integrity of the conducted signal? This condition could be a significant factor because of classroom laboratories and building power supplies. We recommend that a shielded or fiber optic cable be used (eliminating unshielded twisted pair cable).

■ *The number of interconnections of cable to devices*—Will any one cable need to be interconnected to multiple devices? Every run of cable is essentially a point-to-point connection from mainframe to terminal. Therefore, we should not rule out fiber optic cable, even though it is the medium most difficult for making connections.

■ *The weight and size of cables*—Will the weight or size of a particular medium preclude its use in our design? Coaxial cable is typically the largest and heaviest of the conducted media because of its composition and shielding. Fiber optic cable is typically at the other end of the spectrum, being the lightest. Since the largest grouping of terminals/cables is the student cluster of 12 terminals, cable weight and size should not be a factor in determining the choice of conducted media.

■ *The data transmission speed*—Will the transmission speed of the data between mainframe and terminal be so high as to preclude the use of a particular type of cable? Even though shielded twisted pair cable is slower than other media, it is capable of transmitting data for 200 meters at 10 Mbps through relatively benign environments. The current situation should fall within that distance and transmission speed.

■ *The cost of the cable*—Is one medium much more expensive than another? To determine the cost of choosing a particular medium, the total length of all cable runs is calculated below:

6 runs of 50 feet (support personnel terminals)	= 300 feet
12 runs of 250 feet (student terminals)	= 3000 feet
7 runs of 300 feet (faculty terminals)	= 2100 feet
TOTAL	5400 feet

If the approximate cost of shielded twisted pair cable is $0.24 per foot, if coaxial cable is $0.52 per foot, and if fiber optic cable is $0.76 per foot, the cost of 5400 feet of cable is $1296.00, $2808.00, and $4104.00, respectively.

■ *Maintenance of the cable*—Will one type of cable require less or more maintenance than another type of cable? Since the cables will run within building walls and ceilings and through underground tunnels, cable maintenance should be simpler than if the cables were exposed to the outdoor elements. Therefore, cables with a low maintenance requirement should not be necessary.

■ *Conclusion:* The computing center chose shielded twisted pair cable to connect the mainframe computer and the 30 terminals.

■ *Reasoning:* The cable runs are relatively short, the data transfer speeds are within the capabilities of shielded twisted pair cable, the shielding is necessary to protect the signal from environmental noise, and the pricing is reasonable.

SUMMARY

The world of data communications would not exist if there were neither conducted media nor radiated media by which to transfer data. The conducted media consist of twisted pair cable, coaxial cable, and fiber optic cable. The radiated media consist of satellite and radio transmissions. Each of the media have their advantages, disadvantages, and design limitations. Before choosing one medium over another for a given application, you should consider several media selection criteria, such as cost, speed, and maintenance.

All data transmitted over a communications medium is either digital or analog. It is transmitted with a signal which, like data, can be either digital or analog. All signals consist of three basic components: amplitude, frequency, and phase. Two important factors affecting the transfer of a signal over a medium are attenuation (the logarithmic loss of a signal through a conducted medium) and the signal-to-noise ratio.

Since both data and signals can be either digital or analog, four combinations of data and signals can be produced. Digital data with digital signals are represented by digital encoding formats, including the popular differential codes and the newer 4B/5B code. For digital data with analog signals to be transmitted, the digital data must be modulated onto an analog signal. There are three basic forms of modulation: amplitude, frequency, and phase modulation. For conveying analog data with digital signals, two common techniques of conversion are popular: pulse code modulation, which converts analog samples of the analog data to digital; and delta modulation, which tracks analog data, transmitting only a 1 or 0 depending on whether the data rises or falls within the next time period. To transmit analog data over an analog medium, the data is often modulated to another frequency to achieve efficient transmission.

Unfortunately, signals can experience many errors during transmission. Several of the more commonly found errors include phase jitter, echo, crosstalk, white noise, and impulse noise. With proper preventive measures, however, many of these errors can be reduced to insignificant levels.

EXERCISES

1. Using a bar chart, graph the relative costs of twisted pair cable, coaxial cable, fiber-optic cable, satellite microwave, and terrestrial microwave. The *Y*-axis should be cost and the *X*-axis should be the five different media. (Careful: What "costs" are you graphing?) Create a second bar chart graphing the transmission distances of the five media.

2. Given the following—signal frequency = 10,000 Hz, signal power = 5000 watts, and noise power = 230 watts—calculate the data transfer rate using Shannon's formula.

3. If a signal starts at point A with a strength of 2000 watts and ends at point B with a strength of 400 watts, what is the decibel loss of the signal?

4. If a signal experiences a 10-decibel loss over a given section of coaxial cable and the signal began at 50 watts, what is the final power level in watts?

5. If a signal, during the course of transmission, loses one-half of its power, what is the decibel loss?

6. If an amplifier takes an incoming signal and doubles its power, what is the decibel gain?

7. Given the layout of cities, cables, and amplifiers (shown in the graphic for Exercise 7), what is the total decibel gain (or loss) of the entire circuit?

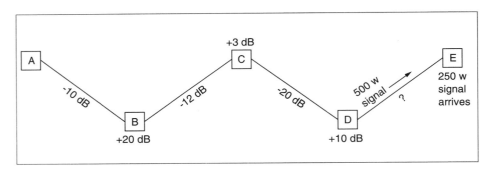

8. If the signal-to-noise ratio for a particular system is 84 dB and the signal power is 2000 watts, what is the corresponding noise power in watts?

9. Draw in chart form (as shown in Figure 2.10) the voltage representation of the bit pattern 11010010 for the digital encoding schemes NRZ-L, NRZ-I, RZ, and Differential Manchester.

10. Given the bit string 1101 1010 0011 0001 1000 1001, show the equivalent 4B/5B code.

11. State both the advantages and disadvantages of pulse code modulation and delta modulation.

12. List all the types of noise discussed in the chapter. Which are reasonably constant and which are not?

13. Show the sequence of bits transmitted when the Baudot code is used to represent the character string ABC12D56.

14. List four possible uses for the eighth bit of the ASCII character set.

15. Given the bit string 00110101, show the equivalent analog sine-wave pattern using amplitude modulation, frequency modulation, and phase modulation.

16. In quadrature phase modulation, why couldn't the four phase changes be 90-, 180-, 270-, and 360-degree phase changes?

17. *Solve It Yourself Case Study:* You are asked to recommend the type of wiring for a manufacturing plant. The size of the plant is 200 meters by 600 meters and houses large, heavy machinery. There are over 200 devices in this plant that must be connected to a computer system. Each device transmits data at 2 Mbps and sends a packet of data every 2–3 seconds. Any cable recommended has to be suspended high overhead in a hard-to-reach conduit system. What type of cable would you recommend? In arriving at your answer, follow the same guidelines as those given in the case study.

CHAPTER 3

THE PHYSICAL LAYER: PART TWO

INTRODUCTION

Chapter 2 began our discussion of the lowest layer of the OSI model: the physical layer. We discussed the basic elements of data, signals, data codes, and types of errors. This chapter continues the discussion of the physical layer by introducing multiplexing, terminals, modems, and interface standards.

MULTIPLEXING

Under the simplest of conditions, a medium is capable of carrying only a single signal. For example, the cable (twisted pair wire) that connects a keyboard to a microcomputer carries a single digital signal. Many times, however, it is advantageous for a medium to carry multiple signals at the same time. The technique of transmitting multiple signals over a single medium is **multiplexing**.

For multiple signals to share one medium, the medium is somehow divided up among the signals. Presently there are two basic ways to divide a medium: a division of frequencies, or a division of time. In frequency division, nonoverlapping frequency ranges are assigned to the different signals. In the previous chapter we saw a simple example of frequency division involving the telephone system (Figure 2.13). The dialogue (signal) of one user was assigned one set of frequencies for transmission while the dialogue (signal) of the second user was assigned a second set of frequencies.

Time division requires that the available time of a medium be divided among the users. As a simple example, suppose two users A and B wish to transmit their data over a shared medium to a distant computer. If we allow User A to transmit during the first second, then User B during the next second, followed again by User A during the third second, and so on, we have created crude time division multiplexing. We examine each of these multiplexing techniques in more detail in the following sections.

Frequency Division Multiplexing

Frequency division multiplexing (FDM) is the assignment of nonoverlapping frequency ranges to each user of a medium. The **multiplexor** is the device that accepts the input from the multiple users, assigns frequencies, and transmits the combined signals over a reasonably high-quality medium. A second multiplexor (or demultiplexor) is attached to the receiving end of the medium and reverses the above procedure. Figure 3.1 shows a simplified diagram of frequency division multiplexing with the attached hardware.

A commonly found example of frequency division multiplexing is the coaxial cable that delivers cable television into homes and businesses. Each television channel (or user) is assigned a unique range of frequencies, as shown in Table 3.1. The television set, or the cable TV decoder box attached to the set, is the demultiplexor that separates one channel from the next and presents them as individual data streams to you, the viewer.

Frequency division multiplexing is the oldest multiplexing technique and one of the simplest. To construct an electronic device that separates one range of frequencies from another is not difficult. Typically it is done with a series of electronic filters that allow a given range of frequencies to pass through while blocking all other frequencies.

Frequency division multiplexing is not the most efficient multiplexing technique, however. The medium that carries the multiplexed signals must be capable of supporting the total range of all involved frequencies. If one channel is not used, that frequency range is still assigned to that one channel and is thus wasted. (Dynamic assignment of frequency ranges does exist, but that is an advanced and expensive technique and is not discussed here.)

Synchronous Time Division Multiplexing

Time division multiplexing (TDM) is divided into two techniques: **synchronous** TDM, and **statistical** or intelligent TDM. Synchronous time division multiplexing is the simpler design but suffers the same inefficiency of unused space as frequency division multiplexing. Given n inputs, a synchronous time division multiplexor accepts one

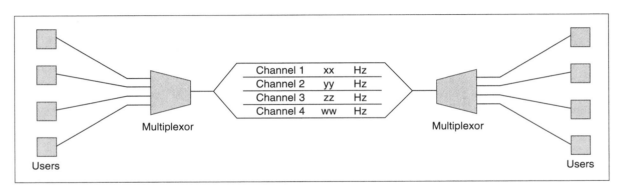

FIGURE 3.1   Simplified example of frequency division multiplexing

TABLE 3.1   Assignment of Frequencies for Cable TV channels

Channel	Frequency in MHz	Channel	Frequency in MHz
Low-Band		High-Band	
2	54-60	7	174–180
3	60-66	8	180-186
4	66-72	9	186-192
5	76-82	10	192-198
6	82-88	11	198-204
		12	204-210
Mid-Band		13	210-216
A	120-126		
B	126-132	Super-Band	
C	132-138	J	216-222
D	138-144	K	222-228
E	144-150	L	228-234
F	150-156	M	234-240
G	156-162	N	240-246
H	162-168	O	246-252
I	168-174	P	252-258
		Q	258-264
		R	264-270
		S	270-276
		T	276-282
		U	282-288
		V	288-294

bit from the first device, transmits it over a high-speed link, accepts one bit from the second device, transmits it over the high-speed link, and continues this process until a bit is accepted from the nth device. After the nth device's bit is transmitted, the multiplexor returns to the first device and continues in this simple round-robin fashion. Bytes (octets) instead of bits may be the unit input from each device. The resulting output stream from the multiplexor is shown in Figure 3.2. It is the responsibility of the demultiplexor on the receiving end of the high-speed link to disassemble the incoming bit stream and deliver each bit to its respective owner.

Under normal circumstances, the synchronous time division multiplexor maintains a simple round-robin sampling order of the input devices. Since the high-speed output data stream does not contain any addressing of individual bits, a precise order must be maintained so that the demultiplexor can disassemble and deliver the bits to the respective owners in the same sequence in which they were input.

If a device has nothing to transmit, the multiplexor must still allocate a slot for that bit (byte) in the high-speed output stream. The demultiplexor cannot know which bit is missing if each device does not contain some form of identification. Thus, if only one device is transmitting, the multiplexor still transmits a bit from each input device (Figure 3.3).

The high-speed link between multiplexor and demultiplexor must always be capable of carrying the total of all possible incoming links, even if some of the incoming links

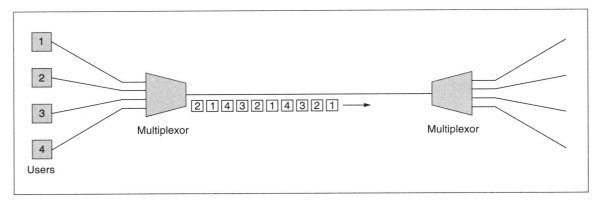

FIGURE 3.2   Sample output stream generated by a synchronous time division multiplexor

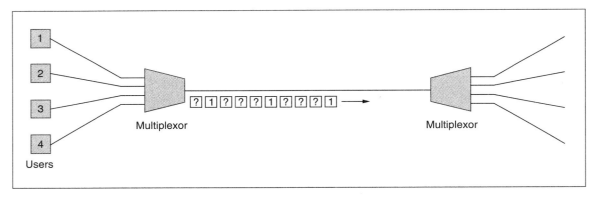

FIGURE 3.3   Multiplexor output stream with only one data source

have no data to transmit. If a multiplexor has eight input sources transmitting data at 9600 bps each, the high-speed line leaving the multiplexor must be capable of transmitting data at 76,800 bps ($8 \times 9,600$) (Figure 3.4).

Statistical Time Division Multiplexing

Both frequency division multiplexing and synchronous time division multiplexing waste unused transmission space; is there a better method? Statistical (or intelligent) time division multiplexing transmits only the data from active users and does not transmit empty time slots, as does synchronous time division multiplexing. For example, if four stations (A, B, C, and D) are connected to a statistical multiplexor but only two of those stations (A and C) are currently transmitting, the statistical multiplexor will transmit only the data from stations A and C, as shown in Figure 3.5. Since only two of the four stations are transmitting, how does the demultiplexor on the receiving end recognize the owners of the data? Some type of address must be included with each octet of data so that the receiver can correctly identify each piece of data (Figure 3.6).

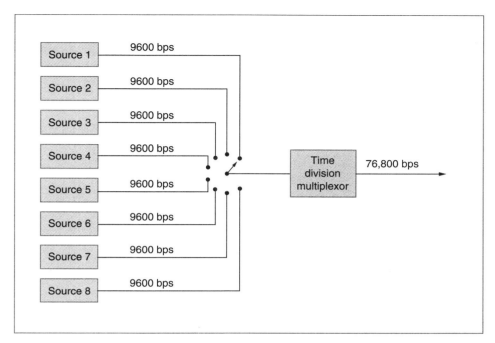

FIGURE 3.4   Time division multiplexor with eight digital input sources

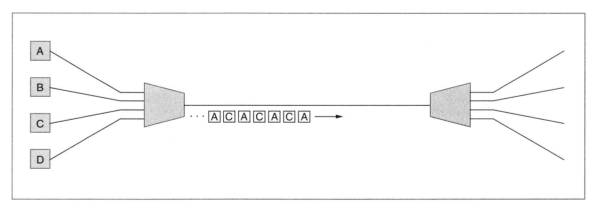

FIGURE 3.5   Two stations out of four transmitting via a statistical multiplexor

An alternative form of address and data is shown in Figure 3.7. In this example, a data block larger than one octet is transmitted. To accomplish this, a length field defining the length of the data block is included along with the address and data. This packet of address/length/data/address/length/data . . . is then packaged into a larger unit by the statistical multiplexor (Figure 3.8). The frame check sequence (FCS) field is the error detection scheme for the remainder of the data block. This particular layout of information fields is actually a creation of the data link layer and is discussed in detail in Chapter 4.

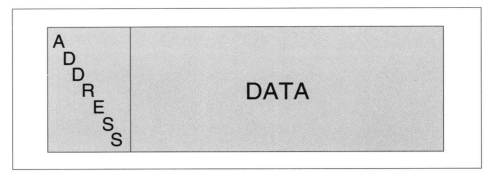

FIGURE 3.6   Sample address and data in a statistical multiplexor output stream

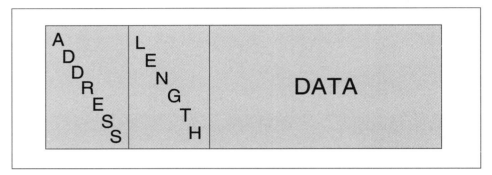

FIGURE 3.7   Alternative address and data fields in a statistical multiplexor output stream

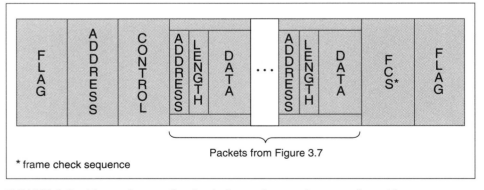

FIGURE 3.8   Frame layout for the information packet transferred between statistical multiplexors

T-1 Multiplexor

In 1984, AT&T presented the **T-1** service, which multiplexed digital data and digital voice onto a high-speed line with a data rate of 1.544 megabits per second. The T-1 multiplexor can divide a single medium into 24 separate digitized voice channels of 64 Kbps each and differentiate between digitized data and digitized voice. For more information on the T-1 multiplexor, see the Further Readings list at the end of the book.

TERMINALS

Before the data can be sent through a medium it must be entered through some device, most often a terminal. A terminal can be defined as an input/output device, typically consisting of a keyboard and screen, that is connected to a local or remote computer on which it is dependent for either computation or data or both. A device does not need to have a keyboard, however, to be classified as a terminal. Sometimes a numeric keypad or the buttons on a push-button telephone are sufficient to provide input to a system. Also, the screen may be nothing more than a one-line display of numbers or characters. Thus, many different kinds of input and output devices can be labeled *terminal*.

Terminal Applications

Applications of terminals are just as varied as the definition. One of the more common applications is as a data entry/display device for mainframe connections (Figure 3.9). Computer programmers, data entry operators, and online transaction systems such as banking and airlines all use terminals as simple data entry and data display devices.

A microcomputer with its keyboard and screen, if it is connected to a host computer system, may also be classified as a terminal. This arrangement gives you the best of both worlds: You have a stand-alone microcomputer with its software and availability, and you have a terminal connection to a host computer with access to its resources.

A third large use of terminals is the **point-of-sale terminal** or cash register found in many retail stores. At one time only a machine for calculating the purchase amount due, this device has become a small computer capable of determining sale prices, inquiring about stock quantities, and processing credit card transactions automatically. If you ever doubt the complexity of today's point-of-sale terminals, exchange a gift that was bought at a sale price for a new item that is also on sale but costs more than the original gift, and ask to charge the difference to a credit card. The point-of-sale teminal can calculate this transaction instantaneously.

Other devices also qualify as terminals:

- Automatic teller machines (ATMs) used for banking
- Credit card gasoline pumps that automatically accept your credit card, dispense gasoline, and bill your account
- Pushbutton telephones, which can be used for transferring funds between banking accounts and inquiring about balances.

Dumb, Smart, and Intelligent Terminals

Terminals are also categorized by the relative level of intelligence with which they are associated. For example, terminals that simply accept and display data and immediately transmit data typed on the keyboard are called **dumb terminals**. They have little or no

FIGURE 3.9   Typical data entry/display terminals

internal memory for performing advanced functions and usually do not have an internal microprocessor. They are generally the least expensive and the simplest to operate.

Smart terminals typically have a microprocessor and a significant amount of internal memory; they are capable of performing more advanced functions than just simple data entry and display. Since smart terminals have a microprocessor and memory they are capable of formatting the information that is to be written to the screen. Figure 3.10 shows a simple example. These smart terminals are capable of writing to the screen in designated positions the prompts required for data entry. Once all the prompts have been written to the screen, the cursor moves from position to position prompting the user to enter the required data. Many smart terminals can even perform edit checks on the entered data to verify that it falls within required specifications. Smart terminals are also capable of displaying data in various protected fields on the screen and preventing the user from changing the data in those fields.

Intelligent terminals are often equivalent to stand-alone microcomputers that have a connection to a host computer. They are capable of performing all the functions discussed above as well as all the functions of a microcomputer. The distinction between smart and intelligent terminals is becoming fuzzier, and soon all terminals will have an imbedded microprocessor and some degree of internal memory.

Terminal Interface

Any time you connect one piece of computer equipment to another, you must ask many questions. For the cable that connects the two pieces of equipment, how many individually conducting lengths of wire does it need? What is the shape or configuration of the connector that is on the end of the cable and plugs into the equipment? Are the voltage levels compatible between the two pieces of equipment? What signals are necessary so that one piece of equipment can "talk" to the other? These questions fall under the category of **interfacing**. We discuss this topic in more detail later in the chapter, but connecting a terminal to a host computer also requires us to understand a little about interfacing.

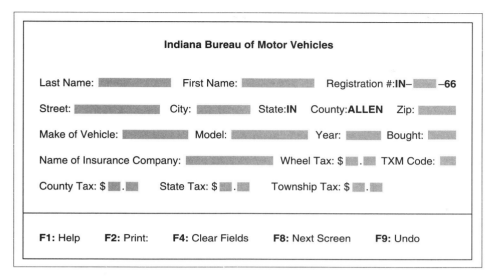

Indiana Bureau of Motor Vehicles

Last Name: ▓▓▓▓▓   First Name: ▓▓▓▓▓   Registration #:**IN**–▓▓▓–**66**

Street: ▓▓▓▓▓   City: ▓▓▓▓   State:**IN**   County:**ALLEN**   Zip: ▓▓▓▓

Make of Vehicle: ▓▓▓▓   Model: ▓▓▓▓▓   Year: ▓▓▓   Bought: ▓▓▓

Name of Insurance Company: ▓▓▓▓▓▓   Wheel Tax: $ ▓▓.▓▓   TXM Code: ▓▓

County Tax: $ ▓▓.▓▓   State Tax: $ ▓▓.▓▓   Township Tax: $ ▓▓.▓▓

F1: Help   **F2:** Print: **F4:** Clear Fields   **F8:** Next Screen   **F9:** Undo

FIGURE 3.10   Example of a formatted screen on a smart terminal

For interfacing terminals to host computers, there are two major classifications: the asynchronous or TTY interface, and the IBM 327x interface. The **TTY (teletypewriter) interface** is essentially an RS-232 type interface (see the section entitled Interface Standards in this chapter). The connector is a multiple-pin connector and the cable is usually a multiple-wire twisted pair. The voltage levels and the signals required follow the specifications listed for the RS-232 standard (or close relative of the RS-232 standard).

The second classification, the **IBM 327x interface**, was created by the IBM corporation to connect their 3270 family of terminals to other IBM equipment. It requires a coaxial cable connection and relies on voltages and signals significantly different from those of the TTY interface. Rather than exploring in detail the differences between the TTY and 327x interfaces, let us simply say here that they are significantly different and anyone purchasing or using terminals and their support hardware should be aware that multiple terminal interface standards do exist, and they may not be readily compatible.

Terminal Configuration

A final way to classify terminals is by the number of terminals that share a connection to the host computer and how they are connected. If there is a direct connection between terminal and host, as shown in Figure 3.11, this is a **point-to-point** connection. If multiple terminals share one connection to the host, as shown in Figure 3.12, this is a **multipoint** connection.

Our discussion of multiplexing has shown that for multiple devices to share a single medium, there must be control procedures to prevent several terminals from sending data to the host at the same time. Adding multiplexors to the connection between terminals and host would be costly and impractical. Another technique must be used to ensure that only one terminal transmits at one time. The technique used harkens back to the old days of computing before microprocessors and cheap memory when terminals were relatively dumb. During this period, host computers were called the **primary** and the

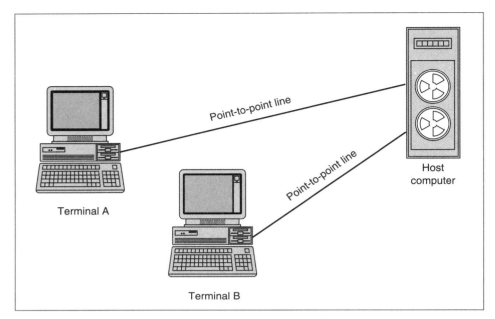

FIGURE 3.11 Point-to-point connection of terminal and host

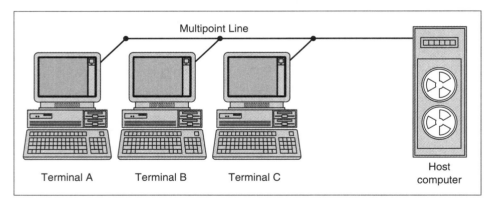

FIGURE 3.12 Multipoint connection of terminals to host computer

terminals were called the **secondaries**. In this technique, a terminal or secondary acted like a well-behaved child and "never spoke unless spoken to." The primary, or host computer, would ask each secondary in turn if it had anything to transmit to the host. This was called **polling**. If three terminals A, B, and C were sharing one connection (Figure 3.13), the primary would begin by polling terminal A. If A had data to send to the host, it would do so. When A had finished transmitting, the primary would poll terminal B. If B had nothing to send, it would inform the primary accordingly and the primary would poll terminal C. When terminal C had finished, the primary would return to terminal A and continue the polling process. This polling technique is termed **roll-call polling**.

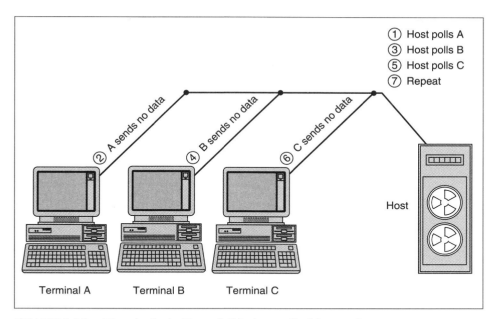

① Host polls A
③ Host polls B
⑤ Host polls C
⑦ Repeat

② A sends no data

④ B sends no data

⑥ C sends no data

Host

Terminal A   Terminal B   Terminal C

FIGURE 3.13   Terminals A, B, and C being polled by a primary

An alternative to roll-call polling is **hub polling**. In hub polling, the primary polls terminal A. When terminal A has finished responding, terminal A passes the poll on to terminal B. When terminal B has finished, it passes the poll on to terminal C. In this fashion the primary does not need to poll each terminal separately. Remember that for the primary to send a poll to a terminal and wait for a response takes time. Under certain circumstances this time may be significant.

What if the primary wishes to send data to a terminal? This process is called **selection**. The primary creates a packet of data with the address of the intended terminal and transmits the packet. Only that terminal (or terminals if the primary wishes to broadcast data to all terminals) recognizes the address and accepts the incoming data.

Point-to-point connection of terminals is clearly superior to multipoint connections with respect to control simplicity. With point-to-point, polling is not necessary since there is only one terminal per line. With multipoint connections, the terminal must possess additional logic to support polling.

Unfortunately, although point-to-point connections use time more efficiently, they require more expensive hardware. If each terminal has a direct connection to the primary, more cabling will be necessary. Also, if two modems are needed to support the connection (one at the primary end and one at the terminal end), a multipoint configuration is able to share the modem among the several terminals (Figure 3.14).

MODEMS

The **modem** (MOdulator/DEModulator) is the device that converts the digital data from a terminal or computer to an analog signal for transmission over an analog

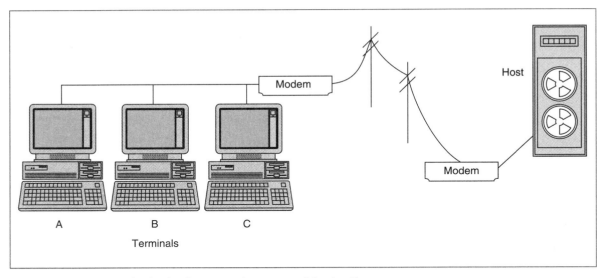

FIGURE 3.14   Terminals sharing a modem on multipoint line

medium (Figure 3.15). While modems appear in many shapes and sizes, they all share a basic set of characteristics:

■ *Transmission speed or rate*—Every modem is designed to transmit data at a certain rate or rates. Many modems can transmit data at a fixed maximum rate but can also drop back to lower rates if the modem on the opposite end of the connection is not capable of supporting the higher rate. Modems may also drop back to a lower data transmission rate if the connection is noisy and data is being corrupted at the higher data rate.

■ *Auto answer*—This feature is the ability of a modem that is operated on the public telephone network to send an answering tone automatically in response to an incoming call. Most modern modems have this feature.

■ *Auto dial*—Auto dial is the ability of a modem to dial calls automatically when commanded to do so by a computer or controller.

■ *Self-testing (loopback)*—Many modern modems are capable of performing two testing functions: local loopback testing and remote loopback testing. In local testing, the A terminal (Figure 3.16) sends data to its local modem, which immediately returns it to terminal A. Remote loopback testing allows the A terminal to transmit data to its local modem, which transmits the signal over the interconnecting medium to the remote modem B. The remote modem B immediately returns the signal to modem A, which returns the data to terminal A.

■ *Half/full-duplex*—Half-duplex is the ability of a pair of modems to transmit data in both directions but in only one direction at a time. Full-duplex is the ability of a pair of modems to transmit data in both directions at the same time. In order to support full-duplex connections, multiplexing is used to divide the medium into two separate channels.

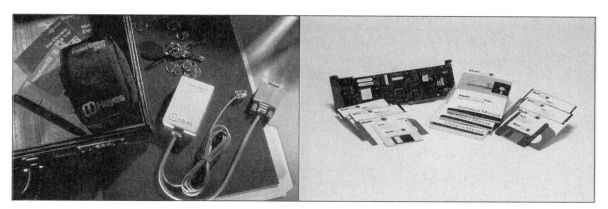

FIGURE 3.15   Examples of common modems; (*left*) external modem, (*right*) internal modem

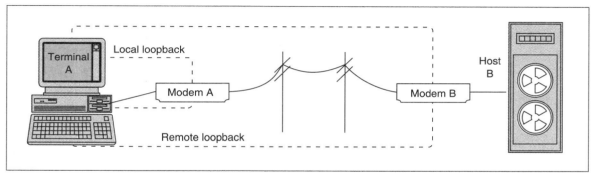

FIGURE 3.16   Example of local loopback testing and remote loopback testing

■ *Data over voice*—This is a frequency division multiplexing technique that combines data and voice over a single twisted pair cable by assigning a portion of the unused bandwidth to the data.

■ *Null modem* (*modem eliminator*)—A null modem is a device (typically a cable) that allows a terminal (or microcomputer) to connect directly to another terminal or microcomputer without the use of intervening modems. Usually a terminal or microcomputer is interfaced to a modem that is connected to a telephone line. If the two microcomputers are close to each other, such as in the same room, there is no need for modems. Since the null modem directly connects the two microcomputers, the cabling requirements are not the same as for a microcomputer directly connected to a modem. The primary cabling change is that the receive data line and the transmit data line are switched. A more complete wiring description of a null modem is shown in Figure 3.17.

■ *Current loop* (*20mA*)—Current loop is an older method of interconnecting Teletype terminals to modems. Logical 0s and 1s are denoted by the absence or presence of current in the connecting cable.

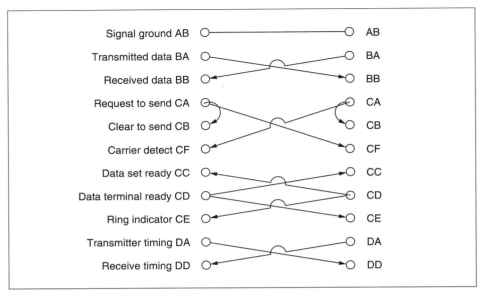

FIGURE 3.17   Wiring schematic of a null modem (modem eliminator)

- *Acoustic coupler*—This is a device that converts electrical signals from the terminal into audio signals, thus allowing data to be transferred over the public telephone system through the handset of the telephone. Acoustic couplers have all but vanished because of their size and the amount of noise they introduce into the circuit.
- *Data compression*—Many modern modems have the ability to compress data, thus increasing the effective amount of data transmitted. Both modems on either end of the transmission line must be capable of the same compression/decompression. Compression techniques are introduced in Chapter 10.
- *Error detection and correction*—Many modern modems can perform an error check on the received data; if an error is detected, the receiving modem informs the sending modem to retransmit the erroneous data packet. Both modems on either end of the transmission line must be capable of the same error detection and correction.

INTERFACE STANDARDS

Many years ago, if one company made a computer terminal and another company made a modem, the odds of the two "talking" to one another were slim. To resolve this problem, various organizations set out to create a standard interface between devices such as terminals and modems. Unfortunately (or perhaps fortunately), many standards have been created over the years. The unfortunate part is the sheer number of standards available and the amount of learning necessary to understand them. The fortunate part is that with so many different transmission and interface environments in existence, one standard alone would not be able to accommodate them all.

The primary organizations involved in making standards are these:

CCITT   Comite Consultatif International Telephonique et Telegraphique
EIA   Electronics Industries Association
IEEE   Institute for Electrical and Electronics Engineers
ISO   International Standards Organization
AT&T/Bell   American Telephone & Telegraph
ANSI   American National Standards Institute

One of the first items standardized was the nomenclature for the devices (such as terminals, modems, and computers) typically found on a communications link. **Data terminal equipment (DTE)** means devices such as terminals and computers. **Data circuit-terminating equipment (DCE)** comprises devices such as modems and multiplexors. Figure 3.18 shows a typical circuit with appropriate DTEs and DCEs labeled. Remember that most of the information discussed in this section deals only with the connection *between* a DTE and a DCE.

An interface standard generally consists of four parts or components: the electrical component, the mechanical component, the functional component, and the procedural component. All the standards that exist address one or more of these components. The electrical component deals with voltages, line capacitance, and other electrical terms. We discuss this component only briefly when we examine the different standards. The mechanical component deals with items such as the connector (or plug) description. The functional component describes the pins/wires/circuits that are used in a particular interface. The procedural component describes how the particular circuits are used to perform an operation. For example, the functional component may describe two circuits: Request to Send, and Clear to Send. The procedural component tells us how those two circuits are used so that the DTE may transfer data to the DCE. As a first example of an interface, we will examine one of the most common standards, EIA-232-D.

EIA-232-D

The EIA-232-D interface standard appeared in 1987 replacing the older RS-232c standard. While the two standards are different, they are similar enough that you should be

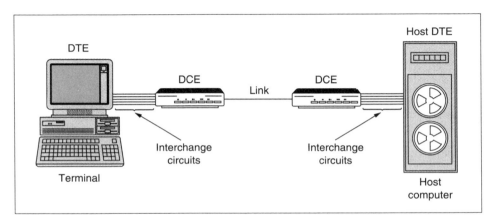

FIGURE 3.18   Typical circuit showing DTEs and DCEs

able to grasp the concepts of RS-232c from a description of EIA-232-D. While the older RS-232c is still found in a majority of installations, the newer EIA-232-D is slowly being accepted. However, many people do not realize there are two standards or what their differences are.

EIA-232-D is an interface standard for connecting a DTE to voice-grade modems (DCE) for use on analog public telecommunications systems using synchronous or asynchronous transmission. (Synchronous and asynchronous transmission are discussed in detail in Chapter 4.) EIA-232-D defines all four components of an interface standard. The first component, the mechanical interface, is defined by another standard, ISO 2110. ISO 2110 precisely defines the size and configuration of a 25-pin connector (called a DB-25 connector). (The older RS-232c did not describe the size and shape of the connector.) Figure 3.19 shows the connector's configuration.

The electrical component of EIA-232-D is based on the standard CCITT V.28, which describes the electrical characteristics for an unbalanced interchange circuit. In an unbalanced interchange circuit, voltage levels are detected at the receiver by the relative voltage difference between the signal circuit and Signal Ground (circuit AB). A voltage difference of more than +3 volts is considered to be ON while a voltage difference of more than –3 volts is considered to be OFF.

The functional and procedural components are taken from the CCITT V.24 standard. CCITT V.24 defines a list of 43 interchange circuits that can be used for an interface design. EIA-232-D uses approximately 20 of those 43 interchange circuits. Those circuits are shown in Table 3.2. Rarely are all circuits used. In fact, it is not uncommon to find a DTE/DCE interface consisting of only three wires: Signal Ground, Transmitted Data, and Received Data. Nonetheless, it is worthwhile to examine in more detail the other circuits and their functions.

Transmitted Data represents the line (wire) on which the DTE sends data to the DCE. *Received Data* represents the line on which the DCE sends data to the DTE. *Request to Send* is sent from the DTE to the DCE when the DTE wishes to transmit data. *Clear to Send* is sent in response to *Request to Send* when the DCE is ready to accept data transmitted from the DTE. *DCE Ready* is a signal sent to the DTE indicating the status of the local DCE. *DTE Ready* is a signal sent to the DCE indicating the ready status of the local terminal. *Received Line Signal Detector* indicates to the DTE that the DCE is receiving a suitable carrier signal and that a telephone connection has been established between the two DCEs. *Transmission Signal Element Timing* and *Receiver Signal Element Timing* provide signal element timing information that can be used to synchronize the transmission and reception of the binary 1s and 0s. The secondary data and control lines are equivalent in function to the primary data and control

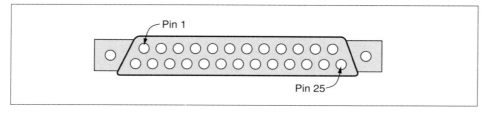

FIGURE 3.19   ISO 2110 Connector

TABLE 3.2 EIA-232-D Interchange Circuits

Pin	Circuit	Source	Description
1	AA	—	Shield
2	BA	DTE	Transmitted Data
3	BB	DCE	Received Data
4	CA	DTE	Request to Send
5	CB	DCE	Clear to Send
6	CC	DCE	DCE Ready
7	AB	—	Signal Ground
8	CF	DCE	Received Line Signal Detector
12	SCF	DCE	Secondary Received Line Signal Detector
13	SCB	DCE	Secondary Clear to Send
14	SBA	DTE	Secondary Transmitted Data
15	DB	DCE	Transmission Signal Element Timing
16	SBB	DCE	Secondary Received Data
17	DD	DCE	Receiver Signal Element Timing
18	LL	—	Local Loopback
19	SCA	DTE	Secondary Request to Send
20	CD	DTE	DTE Ready
21	RL	DCE	Remote Loopback
22	CE	DCE	Ring Indicator
23	CH	DTE	Data Signal Rate Selector
23	CI	DCE	Data Signal Rate Selector
24	DA	DTE	Transmitter Signal Element Timing
25	TM	—	Test Mode

lines, except that they are for use on a reverse or backward channel. The *Local Loopback* signal permits a test on the loop between the DTE and the local DCE. *Remote Loopback* permits a test on the loop from the local DTE, through the local DCE, over the transmission lines to the remote DCE, then back (Figure 3.16). *Ring Indicator* indicates to the DTE that a ringing signal is being received at the local DCE (someone is calling). *Test Mode* indicates whether the local DCE is in a test condition.

EIA-232-D Interface Example

The following two simple examples show the sequence of events that occur during the establishment of a connection between DTEs and DCEs on full-duplex (Figure 3.20) and half-duplex (Figure 3.21) configurations. Please note that the following sequence of events is a greatly simplified version of what really happens and is intended only to demonstrate the basic sequence of an EIA-232-D exchange

In Figure 3.20, the DTE on the left wishes to transmit data to some remote location (the DTE on the right) using a dial-up telephone line. The DTE Ready signal is sent from the DTE to the modem (DCE) when the terminal is operational and is ready to transmit or receive data. The DCE Ready signal is sent from the modem to the DTE when the modem is ready to accept an incoming call or is ready to place an outgoing

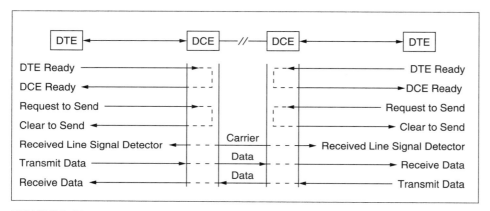

FIGURE 3.20   Full-duplex operations across a DTE/DCE interface

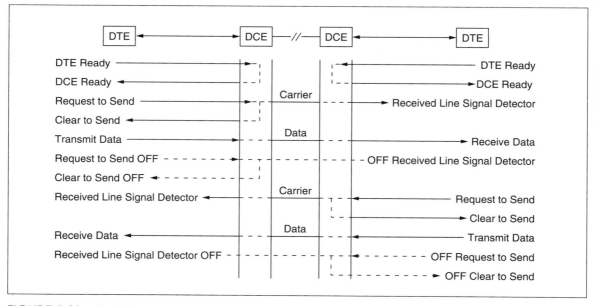

FIGURE 3.21   Half-duplex operations across DTE/DCE interface

call. The terminal, seeing the DTE Ready and DCE Ready signals present, raises the Request to Send signal, informing the modem it wishes to place a call and transmit data. The modem (DCE) responds with a Clear to Send signal. The DTE then transfers the phone number of the station it wishes to call to the DCE, and the DCE dials the phone and places the call. When the remote station's modem answers the phone, a carrier signal is established between the two modems, and each modem informs its DTE that an acceptable signal has been received by raising the Received Line Signal Detector. The DTE can then send data to its modem for transmission over the line or can receive data from the remote station. When the terminal has finished transmitting its data, it drops its Request to Send, causing the modem to drop its Clear to Send signal and ending the phone call with the remote station.

The second example of a half-duplex operation is more involved, since only one end or the other may transmit data over the telephone line at one time. Figure 3.21 roughly demonstrates the sequence of operations between two stations performing a half-duplex data transfer. Once the DTE Ready and DCE Ready signals are raised, the DTE wishing to transmit raises its Request to Send line, which causes its DCE to transmit a carrier signal to the remote DCE. When the carrier signal has been established, a Clear to Send signal is returned to the local DTE and the DTE begins transmitting data. When the DTE has finished sending data, it drops its Request to Send, causing the local DCE to drop its Clear to Send and the carrier signal. The remote DCE notices that the carrier signal has been dropped; its DTE can then raise its Request to Send to transmit a reply.

Comparison Between RS-232c and EIA-232-D

While the older RS-232c and the newer EIA-232-D are essentially equivalent protocols, there are small differences between the two, as the EIA-232-D incorporates CCITT V.24, V.28, and ISO 2110 into the standard. The newer protocol also includes local loopback, remote loopback, and test mode interchange circuits for testing purposes. Protective ground has been redefined and a shield to reduce interference has been added. For a more detailed listing of pin configurations and their changes, the reader is referred to Further Readings at the end of the book.

EIA RS-449

After several years of use, many people realized the shortcomings of RS-232c, particularly the data transmission rates, cable lengths, and ability to perform testing. To address this, EIA in 1975 introduced a new interface standard to replace RS-232c. RS-449 was similar in many ways to RS-232c except for several major improvements, listed below:

1. The electrical component was separated from RS-449 and two new standards (RS-422A and RS-423A) were created, addressing only the electrical component.
2. Ten additional circuits were added, including loopback testing.
3. The connector was changed from a 25-pin to a 37-pin connector.

The electrical component defined by RS-422A (X.27) and RS-423A (X.26) provides for a balanced transmission or an unbalanced transmission, respectively. The unbalanced transmission of RS-423A (essentially a signal wire plus a ground wire) provides for a data transmission between DTE and DCE of 3 Kbps at 1000 meters and 300 Kbps at 10 meters. The balanced transmission of RS-422A (essentially two wires—one for the signal and one for a voltage reference) provides for a data transmission between DTE and DCE at 100 Kbps at 1200 meters and 10 Mbps at 12 meters. The balanced transmission is superior to the unbalanced since the unbalanced transmission relies on a ground, and the grounds from one location to another can be different, producing a voltage difference.

Other V-Series Recommendations

Following is a list of the more commonly found CCITT V-series recommendations. There is no need to memorize the list (unless you are extremely bored). It is provided mainly as a reference for future needs, but you should note the popularity of several of the standards, including V.22, V.22bis, V.32, V.32bis, V.42, and V.42bis. (Bis is interpreted as a "secondary" standard.)

V.1	Gives brief description of terminology pertaining to binary symbols and signals.
V.2	Provides direction on permissible power levels to be used on equipment.
V.5	Describes signaling rates (bps) over half-duplex or duplex dial-up links.
V.6	Describes signaling rates (bps) over dedicated links.
V.7	Defines terms concerning data communications over the telephone network.
V.10	Defines the electrical characteristics of an unbalanced interchange circuit. Equivalent to RS-423A and X.26.
V.11	Defines the electrical characteristics of a balanced interchange circuit. Equivalent to RS-422A and X.27.
V.21	Defines an interface standard for 300 bps modems using asynchronous or synchronous transmission and a form of frequency modulation.
V.22	Defines the interface standard for modems transmitting data at 1200 bps over full-duplex, asynchronous or synchronous, two-wire leased or dial-up lines. Employs phase shift modulation.
V.22bis	Is similar to V.22 except that this standard is capable of 2400 bps data transmission speed and uses quadrature amplitude modulation.
V.23	Defines an interface standard for 1200 bps half-duplex transmission using frequency modulation.
V.24	Provides a list of definitions for the interchange circuits.
V.25	Describes the conventions for automatic dial-and-answer interfaces.
V.26ter	Defines an interface standard for 2400 bps half- or full-duplex using asynchronous or synchronous transmission and phase shift modulation. (V.26 full-duplex synchronous, and V.26bis half-duplex synchronous.)
V.27bis	Defines an interface standard for 4800 bps half- or full-duplex synchronous transmission using phase shift modulation.
V.28	Defines the electrical characteristics for an unbalanced interchange circuit.
V.29	Defines an interface standard for 9600 bps half- or full-duplex synchronous transmission using quadrature amplitude modulation.

V.32   Is a very popular interface standard that transmits data at 9600 bps over two-wire dial-up lines or two- or four-wire leased lines. It operates in full-duplex asynchronous or synchronous mode and uses Trellis-encoding modulation to reduce errors. V.32 is quickly becoming one of the most popular data transmission standards.

V.32bis   Is similar to V.32 except for two important modifications: V.32 bis transmits data up to 14,400 bps (with a fallback to four slower data rates), and V.32 bis modems can converse with each other to agree on an acceptable data transfer rate. If the communications line suddenly becomes noisy, the two modems can renegotiate a slower transfer rate.

V.33   Is similar to V.32 but operates only on synchronous, full-duplex four-wire leased lines.

V.35   Defines an interface standard for DTEs/DCEs that interface to a high-speed digital carrier. Transmits data at 48 Kbps using a form of frequency modulation.

V.42   Defines a standard for providing a level of error control between two V.42 modems.

V.42bis   Defines a standard for providing data compression for the information transmitted between two V.42 bis modems. V.42 bis allows a compression ratio of approximately 4:1.

V.50   Sets standard limits for transmission quality for data transmission.

V.54   Is a loop test device for modems.

V.57   Is a comprehensive data test set for high data signaling rates.

X-Series Recommendations

In case you haven't seen enough standards, we list here a few of the X.series recommendations. Some of these standards do not apply directly to the modem interfacing discussion at present and are introduced on their own in later chapters. Please note that the X.21 standard was created as a replacement for the older RS-232c standard.

X.3   Definition of Packet Assembler-Disassembler for public data networks (Chapter 7).

X.20   Interface between DTE and DCE for start/stop transmission services on public data networks.

X.21   A "digital" interface meant to replace RS-232c. Rather than the myriad control lines of RS-232c, X.21 consists of essentially four control lines: T, C, R, and I. By placing particular values on these four lines, a simpler(?) dialogue between DTE and DCE can be established. For example, the sequence of events shown in Table 3.3 can be used to establish a communications session.

X.24   Equivalent to V.24.

X.25   Used as an interface on a public data network (discussed in detail in Chapter 7).

X.26   Equivalent to RS-423A and V.10.

X.27   Equivalent to RS-422A and V.11.

TABLE 3.3   Sequence of Events using X.21 Standard

DTE		DCE		
T	C	R	I	Comment
1	OFF	1	OFF	ready
0	ON	1	OFF	call request
0	ON	++++..	OFF	proceed to select
ASCII addr	ON	++++..	OFF	DTE transmit address
1	ON	c.p.s.	OFF	connection in progress
1	ON	1	ON	ready for data
data	On	data	ON	data transfer

c.p.s. = call progress signals

X.28   Used as an interface on a public data network (Chapter 7).
X.29   Used as an interface on a public data network (Chapter 7).

MNP Standards

The MNP (Microcom Networking Protocol) standards were developed by Microcom Systems, Inc., to provide error correction and data compression techniques for asynchronous modem data transfer. Both ends of the connection must employ the appropriate modems with the same MNP standards to be able to use the error correction and data compression techniques. MNP Levels 1–4 provide error-free data transmission over asynchronous lines by detecting data transmission errors and sending the necessary information to correct the error. MNP Level 5 incorporates levels 1–4 and also includes a data compression algorithm. The algorithm used allows the modems to compress the data by a factor of 2 to 1, thus enabling a user to transmit twice as much data over the same speed transmission line.

Bell Standard Modems

The Bell System of modems became an industry standard years ago when there were few players in the modem market. Since then Bell has provided a set of standards on which other modem companies have modeled their modems. A brief listing of the more common Bell modems follows:

- 103/133 Series—300 bps, frequency shift modulation, full-duplex, two-wire system.
- 202 Series—1800 bps on conditioned leased lines, 1200 bps on dial-up lines, frequency shift modulation, half-duplex on two-wire systems.
- 201 Series—2400 bps, CCITT V.26 phase shift modulation, half-duplex on two-wire systems, full-duplex on four-wire systems.
- 208 Series—4800 bps, CCITT V.27, phase shift modulation, half-duplex on two-wire systems, full-duplex on four-wire systems

- 212A Series—asynchronous 1200 bps with phase shift modulation or 300 bps with frequency shift modulation, synchronous 1200 bps, full-duplex on two-wire system, for 1200 bps uses CCITT V.22
- 209A Series—synchronous 9800 bps leased lines; can perform functions of a simple multiplexor.

Hayes Standard Modems

The Hayes modems developed by Hayes Microcomputer Products became the de facto industry standard for microcomputer systems during the 1980s. Following is a list of the popular Hayes modems and their operating characteristics:

■ Smartmodem 300	300 bps, Bell 103, Frequency shift modulation
■ Smartmodem 1200	1200 bps, Bell 212A, Phase shift modulation
	300 bps, Bell 103, Phase shift modulation
■ Smartmodem 2400	2400 bps, CCITT V.22 bis
	1200 bps, Bell 212A, CCITT V.22
	300 bps, Bell 103
■ Smartmodem 9600	9600 bps, V.32 Full-duplex
	4800 bps, V.32 Half-duplex
	2400 bps, V.22 bis
	1200 bps, Bell 212A, CCITT V.22
	300 bps, Bell 103

CONCENTRATORS

As we saw in an earlier section, multiplexors deal with incoming streams of bits or bytes, and except for statistical multiplexors, they are relatively simple devices. Since they create a unique high-speed output stream of the combined input data, a demultiplexor is necessary to sort out the high-speed stream. A **concentrator**, while it performs basically the same function of combining multiple input streams into one high-speed output stream, is a more complex piece of equipment.

The concentrator is considered a **store-and-forward device**, as it contains the buffer space and logic necessary to accept multiple input streams of varying speeds and produce one high-speed output stream. The output stream from the concentrator looks like a terminal to the host, thus a "de-concentrator" is not needed. While the concentrator is most often placed at the remote location with the terminals, one can also locate a concentrator near the host computer, in which case it is called a **front end processor**.

Concentrators are also capable of performing more advanced functions, such as data compression, forward error correction, device polling, and other network-related functions. In essence, the concentrator is a minicomputer, able to perform many transmission control functions once performed only by host computers.

CASE STUDY

To continue with our case study, the computing center of the hypothetical XYZ University has a new question to ponder. Since the seven faculty terminals are separated from the main computer by a distance of 300 feet, perhaps the computing center should replace the seven cables between the main computer and the faculty terminals with a single, higher-speed cable and two multiplexors located in the computer and faculty terminal rooms (Figure 3.22).

The advantages of using multiplexors are as follows:

- Only one cable is required between mainframe and terminals as opposed to the seven separate cables, saving cable costs.
- Less space is needed in the tunnel between buildings when a single cable is used.

The disadvantages are these:

- The one cable would have to be of high-enough quality to support the combined data rates of the seven terminals.
- The cost of the two multiplexors (outright purchase or lease costs) would have to be absorbed.

In Chapter 2, we calculated the costs of the seven 300-foot twisted pair cables as $1092. If we assume that each terminal transmits data at roughly 19,200 bps, the single replacement cable would have to support 7 × 19,200 or 134,400 bps. This does not seem unreasonable. Recall that as the data rate rises, the quality of the transmission medium will have to increase to support the higher transfer speeds. This may necessitate the use of coaxial cable or fiber optic cable instead of twisted pair.

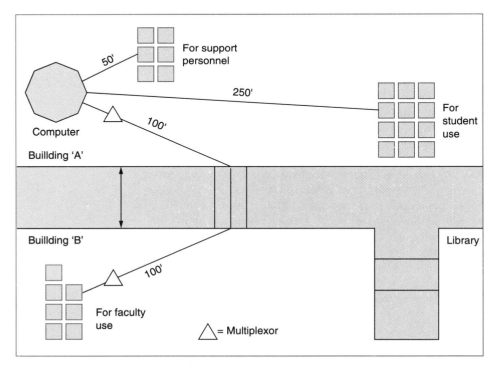

FIGURE 3.22   Hypothetical situation with two multiplexors

Assuming a 300-foot coaxial cable costs approximately $156, replacing the seven twisted pair cables with one coaxial cable results in a savings of $936 (further assuming the seven twisted pair cables have not yet been purchased). An entry level multiplexor capable of supporting a maximum of eight terminals costs approximately $3000. We would need two multiplexors, one for each end of the transmission line, but the cost of the multiplexors does not justify the savings in cable costs. Given the cost of the multiplexors, we would need to replace approximately 35 300-foot cables. Unfortunately, the quoted multiplexors support only eight terminals. A multiplexor that supports 35 terminals would be significantly more than $3000.

Cost alone, however, is not the only criterion. What if there is a very small tunnel or conduit between buildings? It may not be large enough to accommodate multiple cables. A single cable with multiplexors may be the only choice. If future expansion of the number of faculty terminals is envisioned, the simpler and cheaper solution in the long run may be to install one, high-quality cable initially, with upgradeable multiplexors on each end.

To consider a completely different scenario, what if the computing center communicated with the faculty terminal room via seven separate phone lines, one for each terminal? Now, to assess the effective costs of the multiplexors, you have to consider the lease costs of the multiple phone lines. Given the rising costs of leasing phone lines, it may be cheaper to lease or purchase two multiplexors and lease one high-quality phone line as opposed to buying no multiplexors and leasing multiple phone lines.

SUMMARY

For multiple signals to share one medium, the medium must be divided into multiple channels. Presently, there are two basic ways to divide a medium: by dividing frequencies or by dividing time. Frequency division involves assigning nonoverlapping frequency ranges to the different signals. Time division of a medium requires that the available time of a medium be divided among the users. Frequency division multiplexing, synchronous time division multiplexing, and statistical time division multiplexing are the three most commonly found forms of multiplexing.

All terminals possess several basic characteristics and application areas, with a distinction made between dumb, smart, and intelligent terminals. Most terminals have one of two different types of interface: the standard RS-232 interface, or the IBM coaxial interface. You may also distinguish terminals by their interconnection to a mainframe computer. If there is a single connection between mainframe and terminal, it is a point-to-point connection. Terminals that share a connection to a mainframe are in a multipoint or multidrop connection. If the connection is multipoint, some form of polling is used to solicit information from each terminal.

The modem is the device that converts the digital data from a terminal or computer to an analog signal for transmission over an analog medium. Most modems operate according to a set of basic concepts and characteristics and follow one or more interface standards.

While too many interface standards exist to memorize them all, you should understand the main concepts of the common EIA-232-D standard. It defines the four components of an interface standard: electrical, mechanical, functional, and procedural.

Particular emphasis should also be placed on the popular V.22, V.32, and V.42 standards.

EXERCISES

1. State two advantages and two disadvantages of
   a. frequency division multiplexing compared to other multiplexing techniques.
   b. time division multiplexing compared to other multiplexing techniques.
   c. statistical time division multiplexing compared to other multiplexing techniques.
2. When data is transmitted using a statistical multiplexor, the individual units of data must have some form of address that tells the receiver the intended recipient of each piece of data. Instead of assigning absolute addresses to each piece of data, is it possible to incorporate relative addressing? If so, explain its benefits.
3. State the differences between an asynchronous (RS-232) terminal interface and the IBM 327x terminal interface.
4. What is the major difference between a dumb terminal and a smart terminal? Between a smart terminal and an intelligent terminal?
5. Give one example for each of the following: a dumb, a smart, and an intelligent terminal.
6. The hypothetical computing center at XYZ University must decide what type of terminal to purchase and install. Present the advantages and disadvantages of choosing between an RS-232 type terminal and an IBM coaxial type terminal. Present the advantages and disadvantages for the university of a dumb, a smart, or an intelligent terminal.
7. If T_D = time to transmit data
       T_{RP} = time to transmit a roll call poll
       T_{HP} = time to transmit a hub poll
   create two formulas for calculating the approximate total time T_T for n terminals using roll call polling and hub polling.
8. What are the major changes to EIA 232-D compared with the earlier RS-232c?
9. What are the major improvements of EIA RS-449 over the older RS 232c?
10. Many kinds of devices qualify as computer terminals. List as many examples of terminals as possible.
11. Show the sequence of messages sent using both roll call polling and hub polling given three secondaries: A, B, and C. Only C has data to transmit to the primary.
12. State the advantages and the disadvantages of a point-to-point line over a multi-point line.
13. True or False:
    a. All the EIA-232-D interface signals are used only between a DTE and its corresponding DCE.
    b. All the EIA-232-D interface signals are used between the DCE on one end of the connection and the DCE on the other end.
    c. Some of the EIA-232-D interface signals are used between the DCE on one end of the connection and the DCE on the other end.

14. What is the difference between a balanced transmission line and an unbalanced transmission line?

15. The EIA RS-449 standard defines which of the four components of interface standards?

16. What is the baud rate of the V.32 standard? The bit rate? If the baud rate and the bit rate are not the same, explain.

17. *Solve It Yourself Case Study*: Two buildings are separated by a distance of 250 meters. There is a six-inch diameter tunnel connecting the two buildings. The first building has 66 terminals, the second building a mainframe computer. How would you recommend connecting the terminals to the mainframe?

THE DATA LINK LAYER

INTRODUCTION

The second layer of the OSI model is termed the **data link layer** and resides immediately above the physical layer. Recall from Chapters 2 and 3 that the physical layer was responsible for transmitting and receiving the raw data over some type of conducted or radiated medium. The responsibilities of the data link layer involve transforming the physical layer into a "transmission line" that appears free of transmission errors. To accomplish this, the data link layer creates a cohesive unit called a **frame**. A frame typically contains one or more fields that are involved in transforming the physical layer into an error-free transmission line. This frame is passed to the sender's physical layer which then transmits the frame over a medium to the physical layer of the receiver. Into this frame is placed the information necessary to

1. perform error checking of the transmitted data
2. control lost and duplicate frames
3. provide the means to prevent a transmitter from overrunning a receiver with data faster than the receiver can process it

If the data link layer can perform these three functions, it has produced a "transmission line" that is free from common transmission errors.

We will see, however, that there is no single method for achieving these goals. Several "standards," some officially approved and some de facto, have emerged over the years. The choice of a standard depends on many factors. To help educate you about these choices, this chapter introduces the more common standards and their characteristics with their advantages and disadvantages.

In examining the more common standards it often helps to categorize them as much as possible. One way to categorize the data link standards is to distinguish the type of connection that exists between devices: point-to-point connection, multipoint connection, or broadcast connection. Point-to-point and multipoint connections are each further subdivided into two classifications: asynchronous transmission and synchronous transmission. The third interconnection type, the broadcast connection, incorporates most local area networks and radio and satellite systems. We begin our discussion of the data link layer by examining several error detection and error correction techniques, after which we introduce the simpler asynchronous transmission technique. This is followed by several examples of synchronous transmission.

ERROR DETECTION

The transmission of data over an appropriate medium is much like Murphy's Law: if something can go wrong, it will. Even if all possible error prevention techniques were applied before and during the transmission of the data, something will invariably alter the form of the original data, producing new data that is erroneous. Therefore, you must always be prepared to detect errors. Below we examine the more popular error detection techniques, beginning with the simpler methods such as parity checking, and progressing to the more involved methods such as the cyclic checksum.

Parity

Simple parity is the easiest error detection method to incorporate and comes in two basic forms: **even** parity and **odd** parity. The basic concept of parity checking is the addition of another bit to a string of bits to produce either an even number of binary 1s (even parity) or an odd number of binary 1s (odd parity). If we use the seven-bit ASCII character set, we can add a parity bit as the eighth bit.

For example, if we assume even parity and wish to transmit the character 1101011, we would add a parity bit of 1. In this way, our eight-bit stream (11010111) now contains an even number of 1s. If one of the bits is flipped (goes from a 0 to a 1, or vice versa) during transmission, the error can be detected, assuming the receiver understands we are checking for even parity. For example, if we send 11010111 and 01010111 is received, the receiver counts the 1s and sees that there is an odd number, thus indicating an error.

What happens if we send 11010111 (with even parity) and two bits are corrupted? For example, we transmit 11010111 and 00010111 is received. Is an error detected? No, since there is still an even number of 1s. Simple parity can detect only an odd number of bits in error per character. Is it possible for more than one bit in a character to be altered due to transmission error? Unfortunately, yes. It has been shown that isolated single-bit errors occur 50–60% of the time, while error bursts with two erroneous bits (separated by less than 10 nonerroneous bits) occur 10–20% of the time.

Note also that for every 7 bits of data a parity bit is added. The ratio of parity bits to data bits is 1:7. As we shall soon see, this is not a very good ratio.

Double parity (horizontal or longitudinal parity) tries to solve the problem of detecting even numbers of errors by adding additional parity check bits. In this scheme, individual characters are grouped together in a block. Each character (or row) in the block has its own parity bit. Then, after so many characters, a row of parity bits is included. Each parity bit in this last row is a parity check for all the bits in the "column" above. Table 4.1 shows a simple example of double parity.

Note that if 1 bit is altered in row 1, the parity bit at the end of row 1 will signal an error as well as the parity bit for the corresponding column. If 2 bits in row 1 are flipped, the row 1 parity check will not signal an error, but two column parity checks will signal an error. Unfortunately, if 2 bits are flipped in row 1, 2 bits are flipped in row 2, and the errors occur in the same columns, no errors will be detected.

While an extra level of protection is provided by the double parity check, this method also introduces a higher level of redundancy of check bits to data bits. If we are transmitting N characters in a block, the ratio of check bits to data bits is

$N + 8:7N$

To transmit a 20-character block of data, we will need one-fifth as many check bits as data bits.

Arithmetic Checksum

A more advanced technique of error detection is **arithmetic checksum**. While this technique does not have widespread use, it is relatively simple to program and reasonably effective. The checksum value is created by adding together the ordinal values of each character in a transmitted packet. This checksum is then appended to the packet of transmitted data in the form of one or two characters. As a simple example, suppose a transmitter sends the packet of characters "ABC." The ASCII ordinal value for A is 65, B is 66, and C is 67, with a sum of 198. This sum is converted to binary, appended to the data, and transmitted as a fourth character following "ABC." The receiver will accept the three characters of data and the checksum, add the ordinal values of the data, and compare its sum with the checksum transmitted. If the two sums disagree, there has been a transmission error.

If the packet of transmitted data has so many characters that their ordinal values total more than 8 bits (256) or 16 bits (65,536), the checksum becomes too large to represent in one or two appended characters. In this case, some form of truncation of the higher-order bits of the checksum takes place, resulting in either an 8- or 16-bit remainder

Cyclic Redundancy Checksum

The **cyclic redundancy checksum** (**CRC**) is one of the few cases in computer science where you get *more* than you paid for. In all the previous examples of error detection, we

TABLE 4.1   Simple Example of Double Parity

1	1	0	1	0	1	1	1	row 1
1	1	1	1	1	1	1	1	row 2
0	1	0	1	0	1	0	1	row 3
0	0	1	1	0	0	1	1	row 4
0	1	0	0	1	1	1	0	parity row

saw high ratios of check bits to data bits achieving only mediocre error detection. The CRC method typically adds 16 or 32 check bits to potentially large data packets while approaching near 100 percent error detection. Clearly this method deserves a close examination.

The idea underlying the CRC error detection method is to treat the packet of data to be transmitted (message) as a large polynomial. That is, the right-most bit of the data is the x^0 term, the next bit is the x^1 term, and so on. If a bit in the message is 1, the corresponding polynomial term is included. Thus, the data 101001101 is equivalent to the polynomial $x^8 + x^6 + x^3 + x^2 + x^0$. The transmitter takes this message polynomial and "divides" it by a given generating polynomial. Currently several generating polynomials are in widespread use. Dividing the message polynomial by the generating polynomial results in a quotient and a remainder. The quotient is discarded but the remainder in bit form is appended to the end of the original message and this unit is transmitted. The receiver takes the incoming data (original message + remainder) and divides it by the same generating polynomial used by the transmitter. If no errors were introduced into the data during transmission, the resulting division should produce a remainder of zero. If an error was introduced during the transmission, the altered original message + remainder will not divide evenly by the generating polynomial, signaling an error condition.

Suppose you want to transmit the data stream 1010011010. (We'll keep the example data packet small so the calculation doesn't take all day.) For our example, we will use the generating polynomial $x^5 + x^4 + x^2 + 1$. This is not a standard generating polynomial, but one created for this simplified example. To perform the division, we must first append to the end of the original message r 0s, where r is the degree of the generating polynomial. In our example, r is 5, so we append 5 0s. Note also—and this is important—that the division is performed with modulo-2 arithmetic (exclusive-OR). We are dividing binary values, not decimal numbers (Figure 4.1). The remainder is 10110. If the remainder does not contain r digits, insert the appropriate number of nonsignificant 0s to obtain r digits. This remainder is now appended to the end of the original message resulting in

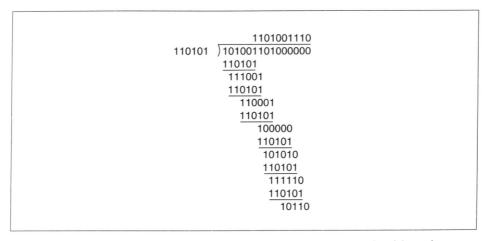

FIGURE 4.1   Example of polynomial division using Modulo-2 arithmetic

the message to be transmitted (1010011010 10110). The receiver will take the incoming message (as it is) and divide it by the same generating polynomial. If there were no transmission errors, a remainder of 0 will result. You should verify this step for practice.

In real life, the transmitter and receiver do not perform longhand polynomial division as we just did. Hardware can now perform the process very quickly. Of course, if hardware can do it, so can software; there are many different algorithms for performing CRC in software. Here we examine the basic hardware that is used to perform the calculation of CRC, using our simple polynomial from above.

The hardware involved is a simple feedback shift register with exclusive-ORs at particular points. Note (Figure 4.2) that where there is a term in the generating polynomial, there is an exclusive-OR between two successive shift boxes. As a data bit is shifted in on the right, all bits shift to the left one position. But before a bit shifts left, if there is an exclusive-OR to shift through, the left-most bit wraps around and exclusive-ORs with the shifting bit. The following example (Figure 4.3) uses the same generating polynomial and message from the longhand division example above. You are encouraged to step through the process to understand the operation.

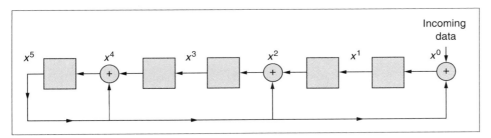

FIGURE 4.2   Hardware shift register used for CRC generation

0 ⊕ 0		0 ⊕ 0		0 ⊕	Incoming data
0	0	0	0	1	1
0	0	0	1	0	0
0	0	1	0	1	1
0	1	0	1	0	0
1	0	1	0	0	0
1	1	1	0	0	1
0	1	1	0	0	1
1	1	0	0	0	0
0	0	1	0	0	1
0	1	0	0	0	0
1	0	0	0	0	0 ⎤
1	0	1	0	1	0 ⎥
1	1	1	1	1	0 ⎬ r 0s
0	1	0	1	1	0 ⎥
Remainder: 1	0	1	1	0	0 ⎦

FIGURE 4.3   Example of CRC generation using shift register method

The CRC method is almost "goof-proof." In fact, the following observations can be made about it. If you choose a generating polynomial having $(x + 1)$ as a factor, and the polynomial has three or more terms, the following detection will result:

Single-bit errors: 100%
Double-bit errors (adjacent or not): 100%
An odd number of bits in error: 100%
An error burst of length less than $(r + 1)$ bits: 100%
An error burst of exactly $(r + 1)$ bits in length:
   $1 - (1/2)^{(r-1)}$ probability of detection
An error burst of length greater than $(r + 1)$ bits:
   $1 - (1/2)^r$ probability of detection.

Four commonly found generating polynomials are:

CRC–12: $x^{12} + x^{11} + x^3 + x^2 + x + 1$
CRC–16: $x^{16} + x^{15} + x^2 + 1$
CRC–CCITT: $x^{16} + x^{12} + x^5 + 1$
CRC–32: $x^{32} + x^{26} + x^{23} + x^{22} + x^{16} + x^{12} + x^{11} + x^{10} + x^8 + x^7 + x^5 + x^4 + x^2 + x + 1$

Error Correction

The concept of **error correction** is often interpreted in two different ways. The major distinction between the two is whether the receiver detects the error and simply informs the transmitter there was an error, or the receiver detects the error and then *corrects* the error. The former case is termed simply *error correction* and has the following scenario: data is transmitted to a receiver; the receiver performs an error detection procedure and discovers the data is corrupted; the receiver returns a message to the transmitter indicating a data transmission error; the transmitter retransmits the original data. Examples of error correction are introduced later in this chapter.

The second interpretation of error correction is **forward error correction** and requires the transmitter to transmit a great deal of additional information along with the original data.

For a data code to perform forward error correction, redundant bits must be added to the original data bits. These will allow a receiver to look at the received data and, if there is an error, recover the original data using a consensus of received bits. To understand these redundant bits, we need to examine the **Hamming distance** of a code, or the smallest number of bits by which character codes differ.

In Figure 4.4, we have created a data code using 3 bits (eight combinations). Each combination is used for a valid code, A through F. In this example, the smallest difference between any two characters is one bit. For instance, the difference between C (011) and D (001) is 1 bit. The Hamming distance of this code is thus 1. When a code has a Hamming distance of 1, no errors can be detected. If we transmit 001 and the middle bit is corrupted causing 011 to be received, how can someone know we sent a D and not a C?

In Figure 4.5, we have created a 3-bit data code but only one-half of the combinations are used. The smallest difference between any two character codes is now 2 bits. The Hamming distance is thus 2. Can we detect single-bit errors? Yes, because if we

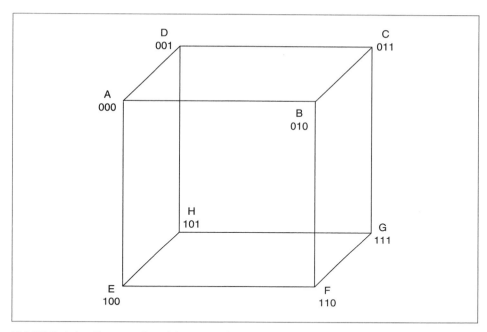

FIGURE 4.4   Data code with Hamming distance 1

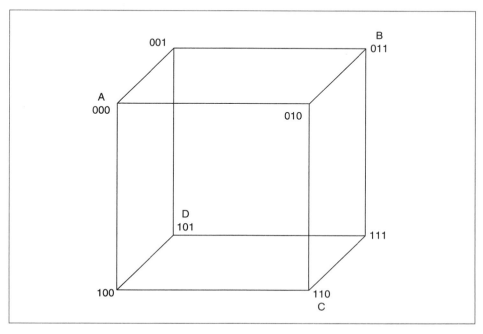

FIGURE 4.5   Data code with Hamming distance 2

transmit, as an example, an A (000) and 001 is received, it can be discarded since 001 is not a valid code. Can we correct single-bit errors? No, since if a 001 is received, the receiver has no idea whether the transmitter sent a 000 (A) or a 011 (B) (or a 101).

Finally, in Figure 4.6 we have created a 3-bit data code using only two of the eight combinations. Now the smallest difference between any two character codes is three bits. The Hamming distance is thus 3. You should verify that we can now detect single- and double-bit errors, and correct single-bit errors.

Clearly, a certain level of redundancy is necessary to achieve forward error correction. A **Hamming code** is a data code that includes the redundant bits necessary to perform forward error correction. A simple example of a Hamming code follows. In this example we have a 4-bit code with the data bits labeled I_3, I_5, I_6, and I_7. Added to these 4 data bits are 3 parity check bits—C_1, C_2 and C_4:

C_1 provides even parity for I_3, I_5, and I_7
C_2 provides even parity for I_3, I_6, and I_7
C_4 provides even parity for I_3, I_6, and I_7

The 7 bits will then be transmitted in the order C_1, C_2, I_3, C_4, I_5, I_6, and I_7.

If you wish to transmit the data 0101, the 7 bits actually transmitted would be 0100101. Note that:

- If data bit I_3 is corrupted during transmission, C_1 and C_2 will detect parity errors.
- If data bit I_5 is corrupted, C_1 and C_4 will detect parity errors.
- If data bit I_6 is corrupted, C_2 and C_4 will detect parity errors.
- If data bit I_7 is corrupted, C_1, C_2, and C_4 will detect parity errors.

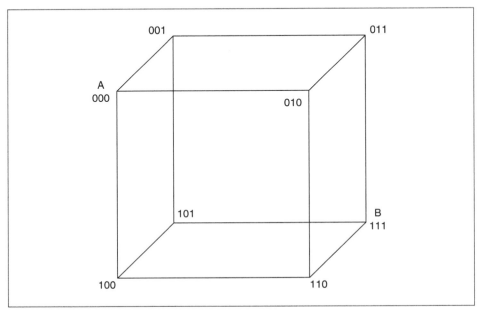

FIGURE 4.6   Data code with Hamming distance 3

- If only C_1 detects a parity error, then it must be C_1 that is in error itself, since there are always at least two C bits that will denote an error for an I bit.

The relationship between C bits and I bits is shown in Figure 4.7.

ASYNCHRONOUS TRANSMISSION

Asynchronous transmission is one of the simplest examples of a data link protocol and is found primarily on point-to-point connections. A single character of data is the unit of transfer between two devices. The transmitting device "prepares" a data character for transmission, transmits that character, then begins preparing the next data character for transmission. An indefinite amount of time may elapse between the transmission of one data character and the transmission of the next character.

To prepare a data character for transmission, you add a few extra bits of information to the data bits of the character creating a small packet of data, called a **frame**. A **start bit** is the first bit of the frame. This bit, which is always a logic 0, informs the receiver that an incoming data character (frame) is arriving and allows the receiver to "lock onto" the following sequence of bits. The end of the frame is marked with one or two **stop bits** (logic 1). While there is usually only one stop bit, some systems still allow a choice of one or two. The start and stop bits have, in essence, provided a "frame" around the data.

Finally, a single parity bit may be added to the data, inserted between the data bits and the stop bit. This parity bit may be either even parity or odd parity and performs a check only on the data bits, not the start and stop bits. Figure 4.8 shows an example of the character A (in ASCII) with one start bit, one stop bit, and even parity. Since each character has its own start/stop/parity, the transmission of multiple characters, such as HELLO, would have the configuration shown in Figure 4.9.

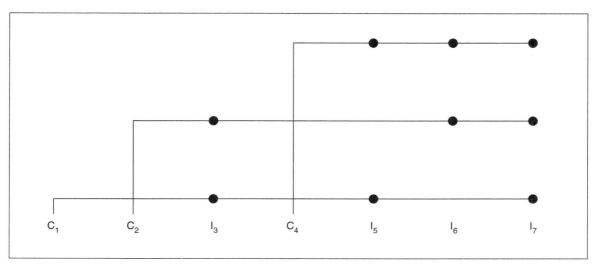

FIGURE 4.7   Relationship of C bits to I bits in sample Hamming code

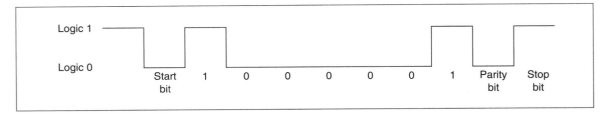

FIGURE 4.8   Example of the character A with start bit, one stop bit, and even parity

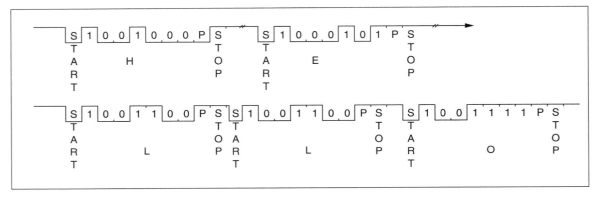

FIGURE 4.9   Example of the character string HELLO with included start, stop and parity bits

The use of asynchronous transmission has advantages and disadvantages. On the positive side, generation of the start, stop, and parity bits is simple and requires little hardware or software. During the early microcomputer years, the simplicity of asynchronous transmission lent itself nicely to the hobby segment of microcomputer-to-host computer links. This association still exists today within the microcomputer hobby market.

The disadvantages of asynchronous transmission are numerous. Because seven data bits (ASCII character code set) are often combined with one start bit, one stop bit, and one parity bit, the resulting transmitted character contains seven data bits and three check bits, for a 7:3 ratio. Almost 50 percent of the transmission is check bits—not very efficient for high amounts of data transfer.

Further disadvantages are that asynchronous transmission provides no way to check for lost or duplicate frames, and no way to stop the sender from overrunning the receiver with too much data. With these strong disadvantages, asynchronous transmission should not be used for any serious amounts of data transfer.

SYNCHRONOUS TRANSMISSION

Synchronous transmission is markedly different from asynchronous transmission in that the unit of transmission is not a single character but a sequence of characters. Much as the start and stop bits framed the data bits in asynchronous transmission, a start sequence and an end sequence frame the sequence of synchronous data. These starting and ending sequences (flags) are typically 8 bits each (an octet) in length but are occasionally larger (multiples of 8).

Usually one or more octets of control information follow the start sequence flag. This octet provides information about the enclosed data or status information pertaining to the sender or receiver or both. For example, if the enclosed information is high-priority data, a unique bit in the control information octet may be set to 1 indicating such a condition. Often the control information contains addressing information telling where the data is coming from or for whom it is intended.

Following the data is almost always some form of error-checking sequence such as the cyclic redundancy checksum. After the error-checking sequence is the end sequence flag. An example block diagram is shown in Figure 4.10.

You may wonder why this procedure is called *synchronous transmission*. To define this term we need first to explain something about clocking signals. Synchronous transmission can take place over many types of interfaces, with the RS232 family being only one example. Recall from Table 3.2 that the EIA-232-D standard included three control lines for providing clocking signals: pin 15, Transmission Signal Element Timing; pin 17, Receiver Signal Element Timing; and pin 24, Transmitter Signal Element Timing. The term *synchronous transmission* derives from the use of these clock signals to synchronize the transmission and reception of the data frame. Note that the synchronization is not between the transmit modem and the receive modem, but between the sending computer and its modem, and between the receiving computer and its modem. Figure 4.11 shows roughly the interconnection between computer and modem.

Any further discussion of clocking signals is beyond the scope of this text. Interested readers should consult the Further Readings section at the end of the book (Sherman 90). Next we examine the most common examples of synchronous transmission.

BISYNC Transmission

Introduced in the mid-1960s, IBM's **BISYNC** protocol was designed to provide a general-purpose data link protocol for point-to-point and multipoint connections. While the protocol itself is dated, the concepts introduced with BISYNC have carried over into many later protocols. We would be remiss if we did not examine the basics of this important standard.

BISYNC (or BSC) is a half-duplex protocol in that it allows data to be transmitted in both directions, but not at the same moment. It may also be termed a **stop and wait**

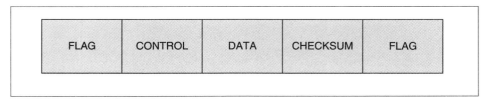

FIGURE 4.10   Block diagram of the parts of a generic synchronous transmission

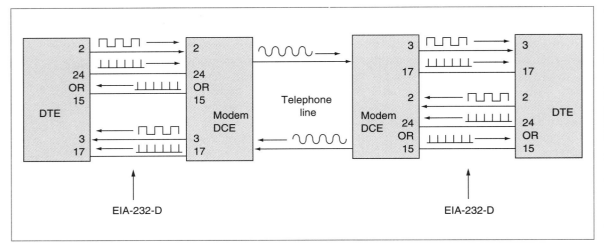

FIGURE 4.11   Interconnection of computer and modem in typical synchronous transmission setup

protocol: One side sends a message, then stops and waits for a reply. Typically, one end of the connection is termed the **primary** while the opposite end is termed the **secondary** (one secondary if point-to-point, multiple secondaries if multipoint). The primary controls the dialogue under most situations and may perform polls and selects of the secondary counterparts.

BISYNC is also termed a **character-oriented** protocol. The messages or frames that are transmitted between primary and secondary are collections of individual characters, many of which are special control characters that drive the dialogue. For example, if the primary wishes to poll the first of multiple secondaries, the following frame is transmitted:

SYN SYN address1 ENQ

The **SYN** character establishes synchronization of the incoming message with the receiver and precedes all message frames. It may also be inserted into the middle of longer messages to maintain synchronization. The **address1** field is the address of the intended secondary. The **ENQ** character, or inquiry, is used to initiate a poll. Note that each control character is an individual character. If you examine the ASCII character chart, you will see that the SYN character is a valid ASCII character with the decimal value 22; the ENQ character is a decimal 5.

If the addressed secondary has nothing to transmit, it responds with

SYN SYN EOT

The **EOT** character signifies the End Of Transmission, or that the secondary has nothing to send.

If the secondary did have something to return to the primary, it would respond with a message such as this:

SYN SYN STX text EOT BCC

The **STX** character signals the Start of TeXt and that data (text) follows. The **BCC** is the Block Check Count (error checksum) appended to the end of the data.

If an optional header with control information is included in the frame, the message would appear like this:

SYN SYN SOH header STX text EOT BCC

where **SOH** indicates the Start Of the Header. On receipt of the data at the primary, if there is no checksum error, the primary responds:

SYN SYN ACK0

The **ACK0** character is a positive acknowledgment for even-sequenced frames. If the frame is corrupted during transmission and the primary receives a garbled message, it replies with

SYN SYN NAK

The **NAK** character is a negative acknowledgment. When the secondary receives the message, it is responsible for retransmitting it.

The more commonly found BISYNC control codes and their meanings follow:

ACK0—**ACK**nowledgment: Positive acknowledgment to even-sequenced frames of data or as a positive response to a select or bid.

ACK1—**ACK**nowledgement: Positive acknowledgment to odd-sequenced frames of data.

DLE—**D**ata **L**ink **E**scape: used to implement transparency (described below).

ENQ—Inquiry: used to initiate a poll message or select message; used to bid for the line in contention mode; used to ask that a response to a previous transmission be resent.

ETB—**E**nd of **T**ransmission **B**lock: used to signal the end of transmission of a block of data. A single message, due to its length, may be broken into multiple blocks.

ETX—**E**nd of **TeX**t: used to signal the end of a complete text message. If a message is broken into multiple blocks, the ETX is placed only at the end of the message.

EOT—**E**nd **O**f **T**ransmission: used to signal the end of a transmission. If the transmission of data consists of multiple messages, the EOT is placed only at the end of the last message. May also be used as a response to a poll.

ITB—end of **I**ntermediate **T**ransmission **B**lock: used to signal the end of an intermediate transmission block.

NAK—**N**egative **A**c**K**nowledgement: used to indicate that the previously received block was in error; used as a negative response to a poll message.

RVI—**R**e**V**erse **I**nterrupt: if a station that is receiving data from another station must send a high-priority message to the sender, the receiver issues an RVI to interrupt the sender and seize control of the line.

SOH—**S**tart **O**f **H**eader: used to indicate an optional header field follows immediately.

STX—**S**tart of **TeX**t: used to indicate the start of text and that the data follows immediately.

SYN—**SYN**chronization: used to provide synchronization as previously described.

TTD—**T**emporary **T**ext **D**elay: used by the sending station to indicate that it cannot send data immediately but does not wish to relinquish control of the line.

WACK—**W**ait before transmit positive **ACK**nowledgement: receiving station is acknowledging previous message but is also saying it cannot accept any more messages until further notice.

Transparency

When a sender transmits a message (or frame) to a receiver, how does the receiver know when the incoming message has ended? As we saw above, the sender inserts an ETB at the end of the block of data. The receiver inputs this ETB character and realizes that the end of the block has been reached and that the checksum follows. At the end of the text, the sender inserts an ETX and once again the receiver responds accordingly. Since the receiver is carefully monitoring the incoming data watching for an ETB or ETX, what would happen if an ETB or ETX suddenly appeared in the middle of the data? The receiver would erroneously assume that the text and transmission had ended and that the following characters would be the checksum. They are not the checksum, however, and the checksum calculation made with them will be erroneous.

Why would a sender insert an ETB or ETX into the middle of the text, causing all this confusion? Consider the case where a sender is transmitting a core dump of a program. A core dump consists of multiple bytes, each byte taking on the value of hexidecimal "00" to hex "FF." If one of those bytes just happens to have the exact same value as the ordinal value of the ETX character, the receiver will believe it has received the ETX character and act accordingly. You must provide a system that allows the binary equivalent of the ETX character to occur in the text but not be recognized as the control character ETX. Such a technique is termed **transparency**. Generally, a scheme must be provided that will allow any bit sequence to be included within the text, even one that may appear as a control character.

To enable transparency, the **data link escape** character (**DLE**) precedes the STX character. Then, to end the stream of text characters, a DLE ETX would be transmitted, rather than just an ETX. The DLE would also be inserted in front of STX, ETB, ITB, EXT, ENQ, DLE, and SYN control characters.

This seems fairly straightforward until you ask whether the binary equivalent of DLE ETX could also appear as data within the text. It surely could. To solve this problem, an extra DLE is inserted before each DLE *in the text and only within the text*. Thus, if the sender encounters a DLE ETX sequence in the text during transmission, it inserts an extra DLE creating DLE DLE ETX. The receiver will see the DLE DLE ETX and discard the first DLE. Since it encountered two DLEs before an ETX, the receiver knows this is not the control sequence DLE ETX but simply text. The only time a single DLE followed by an ETX should appear is after the end of the text. Figure 4.12 demonstrates a before and after example. Take a few minutes to make sure you understand this concept, since you will encounter transparency again in other forms.

Table 4.2 demonstrates a more complete example of two stations engaged in data transfer using the BISYNC protocol.

To summarize the BISYNC protocol, we could ask how, if at all, the protocol satisfies the four basic requirements of the data link layer? First, BISYNC does package the transmitted data and control signals into a cohesive frame. Second, error checking the

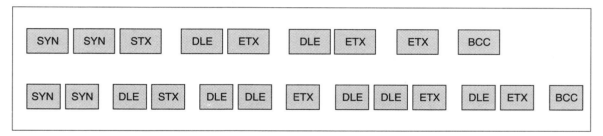

FIGURE 4.12   Example of transparency before and after DLE insertion

frame is accomplished by including a frame check sequence on the end of every packet. Third, lost and duplicate frames are recognized to some degree via the ACK0 and ACK1 fields. However, much like simple parity, you could easily concoct a scenario in which two packets in a row are lost and the system would not recognize it. Finally, flow control can be accomplished by the use of WACKs and ACKs to temporarily suspend transmission by the sender.

Byte Count Synchronous Transmission

The main difference between BISYNC character synchronous transmission and **byte count synchronous transmission** is the inclusion of a length field near the beginning of

TABLE 4.2   Example of a BISYNC Dialogue

Station A		Station B	Description
SYN SYN addr1 ENQ	→		Poll to station 1
	←	SYN SYN EOT	No data to send
SYN SYN addr2 ENQ	→		Poll to station 2
	←	SYN SYN EOT	No data to send
SYN SYN addr3 ENQ	→		Poll to station 3
	←	SYN SYN STX text ETX BCC	Send first part of data
SYN SYN ACK0	→		Acknowledge data
	←	SYN SYN STX text EOT BCC	Send remaining data
SYN SYN ACK1	→		Acknowledge data
SYN SYN addr1 ENQ	→		Poll to station 1
SYN SYN EOT SYN SYN addr3 ENQ	→		Primary selects station 3
	←	SYN SYN ACK0	OK, I'm ready, send the data
SYN SYN STX text EOT BCC	→		Data is sent
	←	SYN SYN NAK	Data garbled
SYN SYN STX text EOT BCC	→		Data sent again
	←	SYN SYN ACK1	Acknowledge data

the frame. This length field is a count of the number of bytes in the remainder of the frame, or just a byte count of the number of characters in the text field portion of the frame. Figure 4.13 shows a simple example of the fields in a byte count synchronous frame.

As the receiver reads in the frame, it encounters the byte count field. It then uses this byte count value to know how much text follows and when to expect the checksum character(s). The inclusion of a byte count field eliminates the need to use a transparent coding system to "hide" control characters within the text. The receiver simply inputs "byte count" characters of text, then automatically reads the checksum. The major disadvantage of using a byte count protocol is the possibility that the receiver will count incorrectly the number of bytes of text being input, and thus read the checksum in the wrong position.

A fairly common example of a byte count synchronous protocol is **Kermit**, the public domain program that allows microcomputer users to transfer files to and from mainframe computers. The basic frame layout of the Kermit protocol is shown is Figure 4.14. Note the length field near the beginning of the frame.

SDLC

All the preceding protocols have been *byte* synchronous protocols in that the receiver examines each byte as it is input, looking for a valid single-byte control code. **SDLC (Synchronous Data Link Control)**, created by IBM in the mid-1970s to replace BISYNC, is a **bit synchronous** protocol; with it, the receiver examines individual bits looking for control information.

There are other major differences between BISYNC and SDLC:

■ SDLC is capable of supporting both half-duplex and full-duplex connections. BISYNC is essentially half-duplex, since the primary sends a message, the sec-

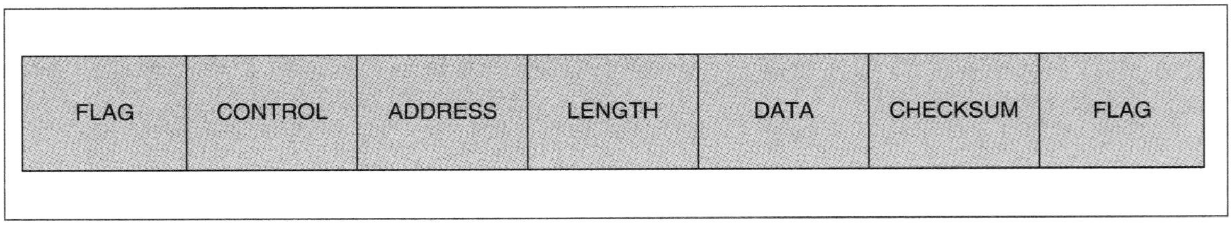

| FLAG | CONTROL | ADDRESS | LENGTH | DATA | CHECKSUM | FLAG |

FIGURE 4.13   Example of a frame using a byte count synchronous protocol

| MARK | LENGTH | SEQUENCE | TYPE | DATA | CHECKSUM |
| 8 bits | 8 bits | 8 bits | 8 bits | $8 \times n$ bits | 8 or 16 bits |

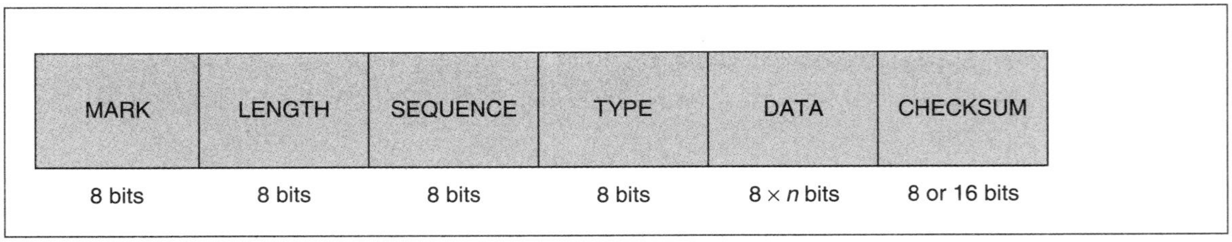

FIGURE 4.14   Sample fields of a message using the Kermit protocol

ondary responds, the primary sends another message, the secondary responds, and so on. SDLC is a full-duplex protocol in that both sender and receiver may transmit at the same time. We will see shortly the mechanism that allows SDLC to support a full-duplex connection.

- SDLC is code independent whereas BISYNC is code dependent. BISYNC requires a data code such as ASCII or EBCDIC that includes control codes such as SOH, STX, or ETX in the code set. SDLC does not rely on character codes to control execution.
- The basic frame format for SDLC differs from that of BISYNC and is shown in Figure 4.15.

The flag field is an 8-bit field with the unique value 01111110. The receiver scans the incoming bit stream looking for the pattern 01111110. When it encounters the pattern, the receiver knows the beginning of the frame has arrived and continues to scan for the remainder of the frame.

Note that the same flag is used to signal the end of the frame. As it looked for the beginning of the frame, the receiver scans the incoming bit stream looking for this unique pattern for the ending. When the ending flag has been encountered, the receiver knows it marks the end of the frame. Again, we must have a way to prevent the bit pattern 01111110 from occurring in the data and CRC fields and signaling the end of the frame erroneously. As the data and CRC portions of the message are transmitted, the transmitter watches for five 1s in a row. If it encounters five 1s, the transmitter automatically inserts a 0. With this procedure, the data and CRC portions of the frame can never contain 01111110. The receiver will also scan the incoming bit stream and if it finds five 1s immediately followed by a 0 within the data and CRC fields of the frame, the extra 0 is discarded. This technique is known as **bit stuffing**.

The address field contains an 8-bit address that is used to identify sender or receiver. In an unbalanced configuration (host computer is the primary, the terminal is the secondary, as with SDLC), the address field always contains the address of the secondary station.

The control field describes the type of frame for this particular message. In SDLC there are three different types of frames:

- I (Information) frames—used to send data, flow control information, and error control information.
- S (Supervisory) frames—used to send flow and error control, but no data.

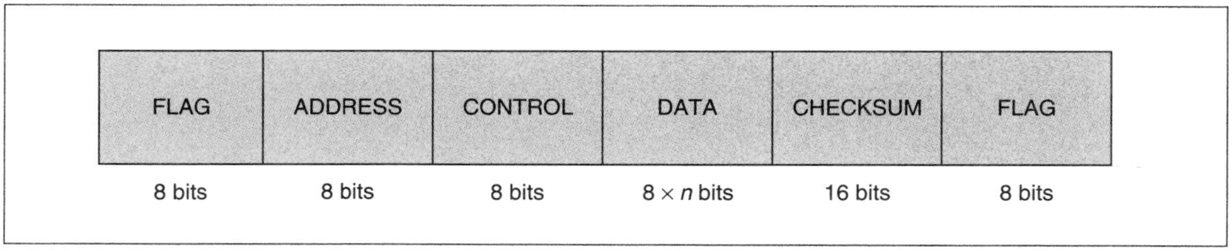

FLAG	ADDRESS	CONTROL	DATA	CHECKSUM	FLAG
8 bits	8 bits	8 bits	$8 \times n$ bits	16 bits	8 bits

FIGURE 4.15   Basic frame format for SDLC

■ U (Unnumbered) frames—used to send supplemental link control information.

Figure 4.16 shows the further breakdown of the control field.

The $N(S)$ and $N(R)$ fields are the send and receive counts for the frames that have been transmitted from and received at a particular station. A station is capable of transmitting frames to another station and receiving frames from another station at the same time. Unlike BISYNC, which sends one frame then awaits a reply, SDLC may send multiple frames to another station before waiting for a reply.

Likewise, a receiving station does not have to acknowledge every single frame. Instead, a receiving station may wait until several frames have arrived before acknowledging any or all of the frames. For example, station A may send seven frames (numbered 0, 1, 2, 3, 4, 5, and 6) to station B. As the frames arrive at station B, station B may wait until the fourth frame (numbered 3) before sending an acknowledgment. Station B would return an acknowledgment to station A with the $N(R)$ count set at 4, implying that frames 0 through 3 were accepted correctly and frame 4 is the *next frame expected*. The $N(R)$ and $N(S)$ counts always reflect the *next* frame number expected or sent. At a later time, station B will acknowledge the remaining three frames by transmitting a frame with the $N(R)$ count set to 7.

So that a transmitting station does not overwhelm a receiving station with a flood of frames, SDLC has a technique that limits the number of frames a station may transmit at one time. More precisely, the **window size** states how many frames may be unacknowledged at any given time. Assume that the window size is 7 and station A sends 6 frames to another station. Station A can still send one more frame since the window size is 7 and only 6 frames have been sent. If the receiving station acknowledges 4 of those frames, station A can then send up to 5 more frames, since 2 of the original 6 frames transmitted have not yet been acknowledged. Because the number of frames that can be transmitted grows and shrinks with transmissions and acknowledgments, the terminology has been more accurately called a **sliding window**.

The S and M subfields of the control field are used by SDLC to further define the type of Supervisory or Unnumbered frames and are discussed shortly.

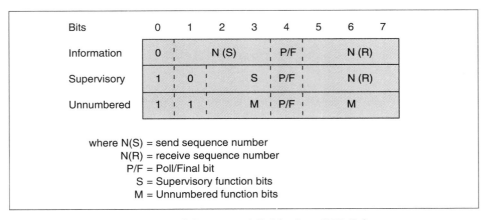

FIGURE 4.16   Bit contents of the control field of an SDLC frame

The P/F bit is the Poll/Final bit. If a primary is polling a secondary, the bit is a poll bit and is set to 1. If a secondary is sending multiple messages to a primary, the last message will have the P/F bit set to 1 and it will act as a final bit.

Following the control field comes the data that is of variable length but always a multiple of eight bits. After the data is the cyclic redundancy checksum, followed by the ending flag (01111110).

The S bits, which exist only within the Supervisory format, are used to specify flow and error control information. Three types of Supervisory messages are available:

- *Receive Ready* (RR) 00: positive acknowledgment; ready to receive an Information frame
- *Receive Not Ready* (RNR) 01: positive acknowledgment, but not ready to receive Information frames
- *Reject* (REJ) 10: negative acknowledgment, go back to the Nth frame (go-back-N) and resend all frames from the Nth frame on.

The M bits, used only within unnumbered frames, represent a command when the frame comes from a primary station, and a response when the frame comes from a secondary station. The available unnumbered frame commands are as follows:

- *Nonsequenced Information* (NSI): M(bits 2 and 3):00–M(bits 5, 6, and 7): 000
- *Set Normal Response Mode* (SNRM): 00–001
- *Disconnect* (DISC): 00–010
- *Optional Response Poll* (ORP): 00–100
- *Set Initialization Mode* (SIM): 10–000
- *Request Station ID* (XID): 11–101
- *Request Task Response* (TEST): 00–111
- *Configure for Test* (CFGR): 10–011

The available unnumbered frame responses are as follows:

- *Nonsequenced Information* (UI): 00–000
- *Nonsequenced Acknowledgment* (UA): 00–110
- *Request for Initialization* (RIM): 10–000
- *Command Reject* (FRMR) 10–100: reject frame, cannot make any sense out of it
- *Request Online* (DM): 11–000
- *Test Response/Beacon* (BCN): 11–111
- *Disconnect Request* (RD) 00–010: request for a disconnect

For a better understanding of SDLC and its commands, we need to examine several examples. In the first example, the primary is polling station A, and station A is requesting initialization information:

$\rightarrow$ FLAG, Address–A,   10–00–1–000, CRC, FLAG   *Primary polls A*

Supervisory Format   Receive Ready   Poll Bit   N(R) = 0

$\leftarrow$ FLAG, Address–A,   11–11–1–000, CRC, FLAG   *A requests online*

Nonsequenced   Request Online   Final Bit
Format

→ FLAG, Address–A,	11–00–1–001, CRC, FLAG	*Primary sets A to normal mode*
Nonsequenced Format	Set Normal Response Mode	Poll Bit
← FLAG, Address–A,	11–00–1–110, CRC, FLAG	*A acknowledges*
Nonsequenced Format	Unnumbered Acknowledge	Final Bit

In the second example, the primary polls station A to learn whether it has data to send, and A replies with three packets of data.

→ FLAG, Address–A, 10–00–1–000, CRC, FLAG   *Primary polls A*

← FLAG, Address–A, 0–000–0–000, data, CRC, FLAG   *A sends first packet*

    data packet #0

← FLAG, Address–A, 0–001–0–000, data, CRC, FLAG   *A sends second packet*

    data packet #1

← FLAG, Address–A, 0–010–1–000, data, CRC, FLAG   *A sends final packet*

    data packet #2, final bit set

→ FLAG, Address-A, 10–00–1–011, CRC, FLAG   *Primary acknowledges packets 0–2*

Primary says packet #3 next expected packet

In the third example, the primary polls station B, and station B replies with several data packets. Due to a transmission error, one packet arrives garbled, and the primary requests retransmission.

→ FLAG, Address–B, 10–00–1–000, CRC, FLAG   *Primary polls B*

← FLAG, Address–B, 0–000–0–000, data, CRC, FLAG   *B sends first packet*

← FLAG, Address–B, 0–001–0–000, data, CRC, FLAG   *B sends second packet*

← FLAG, Address–B, 0–010–0–000, data, CRC, FLAG   *B sends third packet*

← FLAG, Address–B, 0–011–0–000, data, CRC, FLAG   *B sends fourth packet*

← FLAG, Address–B, 0–100–0–000, data, CRC, FLAG   *B sends fifth packet*

→ FLAG, Address–B, 10–00–1–101, CRC, FLAG   *Primary acknowledges packets 000–100*

← FLAG, Address–B, 0–101–0–000, data, CRC, FLAG   *B sends sixth packet*

← FLAG, Address–B, 0–110–0–000, data, CRC, FLAG   *B sends seventh packet, arrives garbled*

← FLAG, Address–B, 0–111–0–000, data, CRC, FLAG   *B sends eighth packet*

← FLAG, Address–B, 0–000–1–000, data, CRC, FLAG   *B sends ninth and final packet*

→ FLAG, Address–B, 10–00–1–110, CRC, FLAG   *Primary acknowledges packets 000–101, requests packets 110 to end*

← FLAG, Address–B, 0–110–0–000, data, CRC, FLAG   *B resends seventh packet*

← FLAG, Address–B, 0–111–0–000, data, CRC, FLAG   *B resends eighth packet*

← FLAG, Address–B, 0–000–1–000, data, CRC, FLAG   *B resends ninth and final packet*

While these examples have been greatly simplified, note that it is quite possible for the primary to carry on conversations with multiple secondaries concurrently.

HDLC

HDLC (High-level Data Link Control) is the data link standard created by ISO that closely resembles SDLC. Typically, anyone dealing with IBM products and software would use SDLC. Note that the two protocols are very similar but not exactly the same. It is possible to make HDLC behave similarly to SDLC, but it may not be possible to make SDLC behave similarly to HDLC. Do *not* assume that the two are interchangeable.

The major differences between HDLC and SDLC are as follows:

■ Extended address—SDLC allows for an 8-bit address, whereas HDLC can be extended to have an address size that is a multiple of 8 bits. To extend the address, a 0 is inserted in the high-order bit position of each octet, except for the last octet of the address which has a 1 in the high-order bit position. Also, in a balanced configuration (available in HDLC but not SDLC), a command frame contains the destination address and a response frame contains the sending address. A balanced configuration involves a primary and secondary, either of which can initiate a data transfer.

■ Extended control field—The control field in HDLC may be 8 bits (as in SDLC) or 16 bits in length. The 16-bit control field allows for 7-bit N(R) and N(S) counts, thus allowing a larger window size for frame transmission.

■ Extended checksum—The cyclic checksum field may be either 16 bits (as in SDLC) or an extended 32-bit checksum (Figure 4.17).

■ Selective reject—Supervisory frames in SDLC can be Receive Ready, Receive Not Ready, and Reject (go back *N*). HDLC adds Selective Reject (SREJ), which tells the sender that a message was in error and to go back to that one message and retransmit it but *not* all the messages that followed it.

■ Additional Unnumbered Commands—Of primary interest here is the fact that multiple modes of dialogue between sender and receiver may be established. The following modes exist in HDLC:

—Set Normal Response Mode (SNRM)—similar to SDLC, primary-secondary type arrangement.

—Set Normal Response Mode Extended (SNRME)—similar to SDLC but control field is 16 bits in length.

—Set Asynchronous Response Mode (SARM)—±the secondary may initiate transmission without explicit permission of the primary, but the primary still retains responsibility for the line (initialization, error recovery, and logical disconnection).

—Set Asynchronous Response Mode Extended (SARME)—same as SARM above but in extended mode (16-bit control field).

—Set Asynchronous Balanced Mode (SABM)—either station may initiate transmission without explicit permission from the other station. No station implicitly retains responsibility. This arrangement is a "peer-to-peer" connection, as opposed to the primary-secondary configuration of SDLC.

—Set Asynchronous Balanced Mode Extended (SABME)—same as SABM above but in extended mode.

If SDLC and HDLC are not acceptable data link protocols, there are other bit synchronous protocols from which to choose. In fact, most major manufacturers of computer equipment and standards organizations seem to have one or two bit synchronous protocols of their own. Some of the more common ones are the following:

- BDLC (Burroughs Data Link Control)
- Honeywell HDLC
- DDCMP (Digital Data Communication Message Protocol) Digital Equipment Corporation
- ADCCP (Advanced Data Communication Control Protocol) ANSI standard
- LAP (Link Access Protocol), CCITT's modified HDLC standard
- LAPB (Link Access Protocol B), CCITT's modified CCITT standard.

Selecting a standard is a lot like selecting a new automobile because there are so many from which to choose. If you don't see a standard you like, you can always wait until next year when all the new models come out.

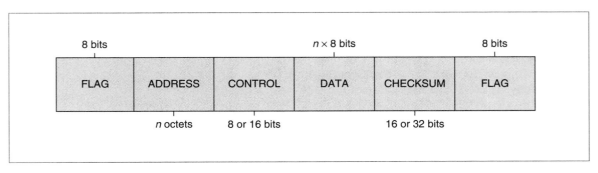

FIGURE 4.17   Basic frame format for HDLC

Comparison of Bit-Synchronous Protocols to Data Link Model

To summarize the bit synchronous protocols SDLC and HDLC, let's once again compare these protocols to the four basic requirements of the data link layer. First, both SDLC and HDLC create a cohesive frame by using the starting and ending 8-bit flag delimiter. Second, error checking is performed on each packet/frame with a cyclic checksum. Third, lost and duplicate frames are controlled by various commands within the protocol. SDLC allows the sender to perform a "go-back-N" retransmission, while HDLC allows both "go-back-N" and "selective repeat." Finally, both SDLC and HDLC perform flow control with the sliding window scheme created by the $N(R)$ and $N(S)$ counts.

MEDIUM ACCESS CONTROL SUBLAYER

While the OSI seven-layer model was designed to support most types of communication systems, it fell short in several areas. One type of communication system that did not fit well into the seven-layer model was broadcast networks. Broadcast networks, such as local area networks and radio/satellite systems, have three characteristics that make them significantly different from other communication systems.

1. One of the main concerns of the network layer of the OSI model is routing a message through the network. This routing often involves deciding which station should receive the message next in order to follow some optimal path. Broadcast networks, as the name implies, broadcast a message from a given point to *all* other stations on the network. Thus, there is generally no need to make any routing decisions. (Some broadcast networks do make routing decisions, as we will see in a later chapter, but this is the exception, not the rule.)

2. There is a close link between the data link layer and the physical layer of most broadcast networks. Because of this relationship, it is difficult to discuss the data link layer without specifying a particular type of hardware. You should recall that the original OSI model was designed to keep each layer separate from the others.

3. Broadcast networks differ from other communication systems in that some type of medium access control procedure is necessary to control who talks when on the one and only medium.

Because of these reasons, the OSI model was modified. Figure 4.18 shows the result of this modification. In this figure, the data link layer has been split into two sublayers: the medium access control (MAC) sublayer, and the logical link control (LLC) sublayer. The MAC sublayer works more closely with the physical layer, thus there is no strictly defined division between the MAC sublayer and the physical layer. The LLC sublayer is primarily responsible for addressing and providing end-to-end error control and end-to-end flow control. You are not expected to understand the LLC sublayer and its duties at this time. Since many of the responsibilities of the network layer do not exist in broadcast networks, the network layer is figuratively smaller (if it exists at all).

To examine in any detail the medium access control layer, we should have a better understanding of local area networks. Because we don't explore these until Chapter 6,

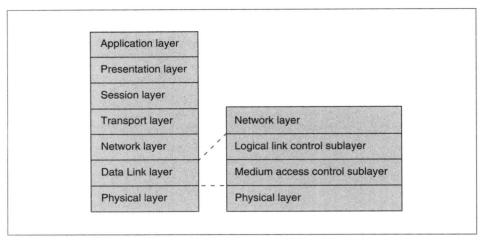

FIGURE 4.18   Division of data link layer into medium access control sublayer and logical link control sublayer

we will defer any more discussion of the MAC sublayer until our deeper examination of local area networks.

CASE STUDY

To apply what we have learned in this chapter, we return to XYZ University. The university computing center is now considering two possible ways to interface its mainframe computer with the outside world. The first idea is to install five modems on the five dial-up ports available on the mainframe. This would allow students, faculty, and staff to dial-in to the mainframe from their home computers. A second idea is to install a modem between the mainframe computer and a leased telephone line to connect XYZ's computer to the large computer network at State University 150 miles away (Figure 4.19).

The idea of adding five dial-up modems raises a couple of interesting questions:

- Should the modems follow an asynchronous protocol or a synchronous protocol?
- What interface standard should the modems incorporate?

As we have seen, when a particular transmission technique or interface standard is selected for a communication line, both ends of the line must follow the same format. Thus, if we choose a particular standard for the dial-up modem on the mainframe computer side, the students who wish to dial in to the mainframe must have modems that follow a similar standard. Therefore, the computing center should choose modems that can operate like the modems the students and the general public can purchase at a retail outlet.

Scanning the current computer advertisements and visiting the local computer retailers, we see that the current trend in modems for home users is quickly

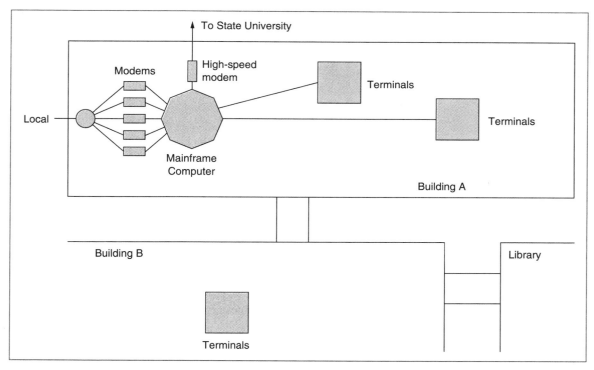

FIGURE 4.19   XYZ University with the proposed additional modems and telephone lines

approaching the level once reserved only for commercial-use modems. Just a few hundred dollars will purchase a modem that supports the newer V.32 and V.42 standards with data compression, error correction, and both synchronous and asynchronous capabilities. Furthermore, most modern modems have the capability to drop back to a slower transmission speed if necessary. Thus, if a student has an older modem that transmits data at a slower rate, the modern modems will recognize this and drop back to accommodate the slower transmission speed.

The second idea—of interfacing XYZ's mainframe with State University's mainframe—can follow different logic, since in that case only XYZ's computing center and State University's computing center need to agree on and purchase a particular type of modem for the two ends of the connection. Since only two modems will be purchased, it is to the advantage of XYZ University to select modems that allow the maximum data transfer for a reasonable cost.

The choice of asynchronous transmission would not be a good one because of the large amount of overhead start, stop, and parity bits that would have to be added to each byte of transmitted data. If the computing center transmits a 2000 byte record using 7-bit ASCII, it would create 14,000 bits of data and 6000 bits of overhead. On the other hand, if a synchronous technique such as HDLC was used to transmit a 2000 byte record, it would create 14,000 bits of data plus only 48 bits of overhead (flag, address, control, CRC [16 bits] and flag). Also, the CRC method of HDLC is vastly superior to the parity check method of asynchronous transmission, and HDLC provides for flow control and selective retransmission of erroneous data packets. After considering these issues, the computing center staff decided on a pair of modems similar to V.32bis modems, which transmit synchronous data at rates up to 14,400 bits per second between the two universities.

SUMMARY

The second layer of the OSI model is the data link layer. A packet of data is passed to the data link layer from the network layer (the next higher layer after the data link layer) and is encapsulated into a frame. A frame is recognized by the header that precedes and the trailer that follows the data packet. Depending on the protocol chosen for the data link layer, various fields of control information are added to the frame to perform error checking, control lost and duplicate frames, and prevent the transmitter from overrunning a receiver with too much data.

One of the most important responsibilities of the data link layer is performing error detection. Techniques used for detecting errors range from the simple parity check to the more complex cyclic redundancy checksum. Forward error correction allows the receiver itself to correct any errors detected.

Asynchronous transmission, one of the simplest data link transmission techniques, simply adds a start bit, stop bit, and parity bit to each single character transmitted. Unfortunately, transmission efficiency suffers because of the large ratio of control bits to data bits. Asynchronous transmission also does not check for lost or duplicate frames, nor does it prevent the transmitter from overrunning a receiver.

Synchronous transmission, characterized by a larger packet of data surrounded by a start flag, stop flag, control field, possible address fields, and checksum field, is a more efficient data link protocol. BISYNC transmission, an older half-duplex synchronous protocol from IBM, creates frames of data from groups of characters. Various control characters are added to the front and rear of the frame to signal operations and inform the sender and receiver of status conditions. Byte-count-oriented protocols simply include a length field within the frame that indicates the number of bytes of data that follow. This length field is used in place of a special control character that signals the end of the data field.

The newer HDLC/SDLC bit-oriented synchronous schemes are full-duplex protocols. They allow primary-secondary conversations, peer-to-peer conversations, and conversations between multiple stations. They are more flexible than the older BISYNC protocol and are not dependent on a particular code set such as ASCII.

The data link layer can be further subdivided into the medium access control sublayer and logical link control sublayer. This division is often associated with local area networks and is discussed in more detail in Chapter 6.

EXERCISES

1. Given the following characters, construct the double (longitudinal) parity check for each character and the block of characters using even parity: 1001010 1110010 0010111 1001101 1111111.

2. Create the arithmetic checksum for the block of characters consisting of the ASCII characters A–I. Will the checksum fit within 8 bits? 16 bits?

3. Calculate the cyclic checksum remainder using the longhand division method for eight 1s followed by eight 0s. Use either CRC-16 or CRC-CCITT.

4. Take the remainder calculated in Exercise 3, append it to the end of the original message, and calculate the cyclic checksum remainder using the longhand division method. If you make no mistakes, the remainder should be zero.

5. Calculate the cyclic checksum remainder using the shift register method for eight 1s followed by eight 0s. Use either CRC-16 or CRC-CCITT.

6. Take the remainder calculated in Exercise 5, append it to the end of the original message, and calculate the cyclic checksum remainder using the shift register method. If you make no mistakes, the remainder should be zero.

7. Given the character string HELLO, show the sequence of start, data, parity, and stop bits that are generated using asynchronous transmission.

8. When a polled BISYNC station has no data to send, what is the response it returns to the primary?

9. What is the BISYNC response to a garbled message received from the primary?

10. You are given the following BISYNC scenario: a primary with three secondaries (A, B, and C); A and C have no data to send to the primary, but station B has three data records it wishes to send.

   a. Show the sequence of BISYNC messages (including polls) that are sent from the primary to the three secondaries.

   b. Show the sequence of BISYNC messages if the second record sent from secondary B is garbled.

   c. Show the sequence of BISYNC messages sent if the ACK from the primary acknowledging the first record sent from secondary B is lost.

11. List as many differences as possible between BISYNC and SDLC.

12. The control field of SDLC describes three different types of frames. What are the three types and what function does each type perform?

13. If a receiver has just received frames 0, 1, and 2, and wishes to acknowledge all of them, what is the $N(R)$ value it returns to the sender?

14. Show the sequence of SDLC frames that are exchanged when the primary polls station C for data, station C returns one frame of data, and the data is not corrupted during transmission.

15. Compare and contrast SDLC with Burrough's BDLC, Honeywell's HDLC, DDCMP, and ADCCP. Each of these will require external readings.

16. What changes were made to the original OSI model to accommodate the newer medium access control sublayer? Why is this newer sublayer associated primarily with broadcast-type networks?

17. Enumerate the differences between SDLC and HDLC.

18. What are the differences between a balanced transmission mode and an unbalanced transmission mode in synchronous transmission?

19. Consider the five most popular modems currently on the market. Create a table in which you compare their characteristics, speeds, features, and costs.

CHAPTER 5

THE
NETWORK
LAYER

INTRODUCTION

The third layer of the OSI model is the network layer. This layer is responsible for transmitting data packets through a computer network in a timely fashion. If there is more than one possible route through the network between the source and destination, the network layer selects the best route based on some type of criteria. The network layer is also responsible for ensuring that the network does not become congested because of computer or line failures. Too many packets in one area of the network or too many packets in the network as a whole will also result in network congestion.

Some networks, because of their design, do not provide a choice of route through the network, and data packet access to the network is strictly controlled. In these cases, it is not unusual to find a very small network layer, if any at all. An example of such a network might be a local area network, which is discussed in detail in Chapter 6.

A final responsibility of the network layer is to pass data between two networks. If the networks are similar, this is not a difficult problem. However, the odds of two computer networks being similar are not high, and the process of passing data between two dissimilar networks can be tedious. This problem is discussed in Chapters 6 and 7.

NETWORK TERMINOLOGY

Before we discuss the two network concepts of routing and congestion, we need to introduce new network terminology; and before we introduce terms which describe the various characteristics of a computer network, we should define computer network. A **computer network** is an interconnection of computers and computer-related equipment used to perform a given function or functions. The types of computers used within a network range from microcomputers through mainframes. The medium of interconnection, as we saw in Chapter 2, can be wire, fiber optic cable, or radiated media. The interconnection can be within a small or local distance, such as a room or building, or as broad or wide as a state, country, or the world. Typical functions of computer networks include the transfer of data between two endpoints, electronic mail services, access to a database of information, and the sharing of computer peripherals and other related computer equipment. Computer networks can assist with specialized operations in numerous fields, such as office production, medicine, navigation, education, law, engineering, and manufacturing, to name only a few.

Underlying the surface of the network is the **subnet**. This substructure is shown in Figure 5.1.

All networks are a collection of at least two basic types of equipment: a station and a node. The **station** is the device that a user interacts with to access a network. It usually contains the software application that allows someone to use the network for a particular purpose. Very often, the station is a microcomputer, or personal computer, but it could also be a terminal, a telephone, or a mainframe computer.

The second device is a **node**. The node allows one or more stations access to the physical network and is a transfer point for passing information through the network. When data or information travels through a network, the information is transferred from node to node through the network. When the information arrives at the proper

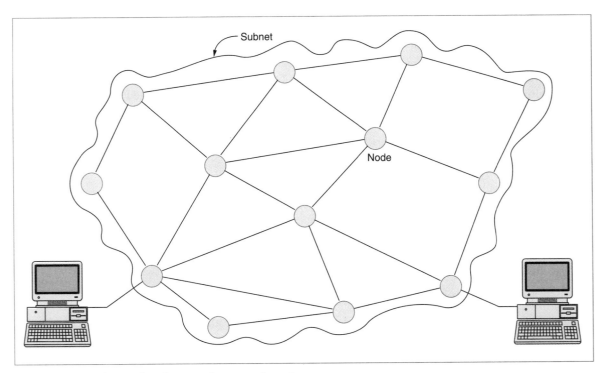

FIGURE 5.1   Network subnet and two end stations

destination, the destination node delivers it to the destination station. This collection of nodes is called the subnet. The type and number of interconnections between nodes and how the network data is passed from node to node is the responsibility of the subnet. It helps to think of the network and the subnet as being almost two separate entities. Clearly, a network would not exist without a subnet, but it should not matter to the network what the subnet looks like. Thus, an application running on a network passes its data to the network (the station), which passes the data to the subnet. The subnet is responsible for getting the data to the proper destination node, which then delivers it to the appropriate destination station.

Three Basic Types of Network Subnets

A network's subnet may be further categorized by the way it transfers information from one end of the subnet to the other. When someone places a call on a dial-up telephone network, a circuit, or path, is established between the person placing the call and the recipient of the call. This physical circuit is unique, or dedicated, to this one call, and exists for the duration of the call. The information (the phone conversation) follows this dedicated path from node to node within the subnet. This is an example of a **circuit-switched** subnet (Figure 5.2).

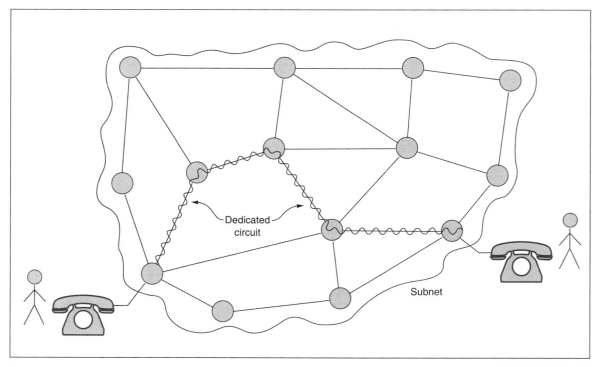

FIGURE 5.2   Two people carrying on a phone conversation using a dedicated path on a circuit-switched network

A second type of network subnet is the **packet-switched** subnet, often found in networks that transfer only computer data (such as the Public Data Network introduced in Chapter 7). In a packet-switched subnet, all data is packaged into fixed-sized packages, called packets. To transmit the data packets across the subnet, no unique, dedicated physical path is established for the duration of the data transfer. Instead, each data packet bounces from node to node as the packets make their way to the destination node. More important, all packets may not necessarily follow the same path as they are transferred from source to destination. As each packet moves through the subnet, each node accepts a packet, and, based on some routing information, passes it on to the next "appropriate" node toward the final destination (Figure 5.3).

Several observations are noteworthy when comparing circuit-switched subnets with packet-switched subnets. In a circuit-switched subnet, all information follows the same path. This path is established when the connection is created and is dedicated for this one and only connection. Even if no information is being passed between the two endpoints, the dedicated circuit still exists. While a circuit-switched subnet is good for large volumes of transmitted information, it can be wasteful when the amount of information transmitted fluctuates.

The packet-switched subnet does not transmit data over unique, dedicated circuits. Since no unique path must be dedicated, the path each packet follows can be dictated

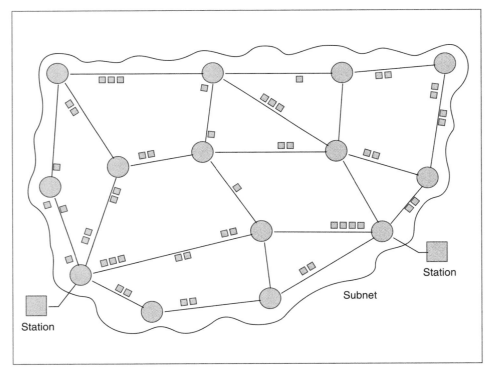

FIGURE 5.3   Multiple packets moving from node to node within a packet-switched network

by the individual nodes within the subnet and it is possible for the packets from many different sources to share the same physical path. Thus, a packet-switched subnet often is more economical than a circuit-switched subnet. As long as the packets arrive at the destination in a timely fashion, does it really matter how they got there? When you mail a letter to a friend across the country, you don't care what cities the letter travels through, as long as the letter arrives in a timely fashion.

A third basic type of subnet is the **broadcast** network. A broadcast network does not concern itself with paths or routes through the subnet because when a packet of data is transmitted through a broadcast network, it goes to *all* other nodes in the network. A common example of a broadcast network is a local area network.

Connection-Oriented Versus Connectionless Networks

Networks may be categorized further by defining the type of connection that may be established on that network. A **connection-oriented** network is one that provides some guarantee that information traveling through the network will not be lost, and that the information packets will be delivered to the intended receiver in the same order in which they were transmitted. Thus, a connection-oriented network provides a **reliable service**. To do so, the network requires that a connection be established between the two end points. If necessary, **connection negotiation** is performed to help establish this connection.

A **connectionless** network is one that does not guarantee the delivery of any information or data. Data may be lost, delayed, or even duplicated. If a large message is broken into multiple smaller messages, or packets, the order of delivery of the packets is not guaranteed. No connection establishment/termination procedures are followed, since there is no need to create a connection. Thus, each packet is sent "as a single entity," not as part of a network connection. Since a connectionless network creates an **unreliable service**, the responsibility of providing a reliable service then falls to the next higher layer, the Transport Layer. That layer is discussed in detail in Chapter 9.

A good analogy for the difference between connection-oriented and connectionless networks is again the difference between the telephone system and the postal service. When you call someone on the telephone, if that person is available for conversation, he or she answers the phone. Once the phone has been answered, a connection is established. The conversation follows, and when one speaker or the other has finished, some sort of ending statement is issued, and both parties hang up. The connection is terminated (Figure 5.4).

If you send a standard letter to someone by the U.S. Postal Service, it will probably be delivered, although there is no guarantee it will be, just as there is no guarantee for when it will arrive. If the letter is lost, you will not know until time passes and you begin to think, "I haven't heard from Bill yet; I wonder if he received my letter?" The unreliable service (no insult meant) offered by the connectionless postal service requires that you must take further actions if you want to ensure that the letter (message) is delivered as intended (Figure 5.5).

A circuit-switched subnet typically supports a connection-oriented network. If the network is connectionless and is going to transmit only a single packet of data, it is

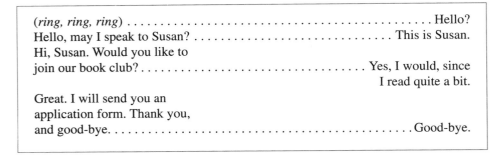

(*ring, ring, ring*) . . . . Hello?
Hello, may I speak to Susan? . . . . This is Susan.
Hi, Susan. Would you like to
join our book club? . . . . Yes, I would, since
I read quite a bit.

Great. I will send you an
application form. Thank you,
and good-bye. . . . . Good-bye.

FIGURE 5.4   Connection-oriented telephone call

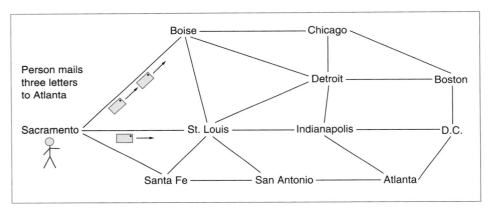

FIGURE 5.5   Connectionless postal network

senseless to have an underlying subnet that is circuit-switched and has to perform all the functions of establishing a dedicated circuit through the subnet.

Broadcast subnets usually support connectionless networks. If someone wishes to transmit data to someone else, the data (usually in the form of a packet) is simply transmitted on the network.

Packet-switched subnets, however, can support either connection-oriented or connectionless networks. To understand how this can be done, we need to examine further the subnet of a packet-switched network.

There are two basic types of packet-switched subnets. A **datagram** subnet is simply a collection of packet-passing nodes. When a data packet arrives at a datagram node, the node examines the destination address of the packet and sends it out the appropriate line based on some routing decision. When the next packet arrives, the datagram node examines this one also, and makes the routing choice all over again. If someone is sending 200 packets from one end of the subnet to the other, the datagram node examines and routes every packet, not realizing that maybe all 200 packets are going to the same place.

The second type of packet-switched subnet is the **virtual circuit**. With this type of subnet, some sort of connection is established before a user transmits any data packets.

As the connection is created, a path through the subnet is chosen and each node along the path is notified that it will be part of a virtual circuit. Each node is then told where to send an arriving packet. Then, any packets that arrive using that virtual circuit will simply be transmitted to the next node in the virtual circuit. One advantage of creating a virtual circuit is the removal of the routing decision-making process from each node in the subnet. This is good if a user of the network plans to transmit many data packets to the same destination. It is more efficient to establish one route at connection establishment time and save each node from having to make a routing decision for every data packet.

One disadvantage of a virtual circuit subnet is its inability to react to network traffic congestion. If a virtual circuit is established, and during transmission of data packets the selected route becomes congested because of heavy traffic or a node or line failure, the virtual circuit cannot react and change the route. The current route has to be terminated and a new route avoiding the congested area has to be created. A datagram node, however, can simply start sending the packets down an alternate route if the original route becomes congested. Thus, the datagram subnet can react quickly to network congestion.

If we now combine connection-oriented and connectionless networks with datagram packet-switched subnets and virtual circuit packet-switched subnets, we have four possible combinations. A connection-oriented network with a packet-switched virtual circuit subnet makes sense, since both require connection and circuit establishment and termination. For the same reasons, it is common to combine a connectionless network with a packet-switched datagram subnet.

What about supporting a connection-oriented network with a packet-switched datagram subnet? While it may sound strange, this combination is not so unusual. The underlying subnet may be datagram, but both ends of the network connection can still create a connection. In this case, the network does not provide reliability by the creation of a virtual circuit; instead, the responsibility of checking for lost or duplicate packets is performed at the endpoints of the network, or maybe in the next layer of the OSI model, the Transport Layer. The final combination, a connectionless network with a packet-switched virtual circuit subnet makes the least sense. It would be wasteful to create a virtual circuit on the subnet if the overall network connection is treated as connectionless, or unreliable. The four combinations of packet-switched networks and subnets is summarized in Figure 5.6.

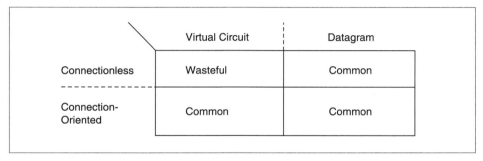

FIGURE 5.6   Summary of four combinations of network connection and packet-switched subnet

ROUTING

We established in the preceding section that a network has an underlying subnet that consists of multiple nodes, each with possible multiple links to other nodes within the subnet. With multiple-linked nodes there may be one or more paths entering as well as leaving a node. If most nodes in the subnet have multiple inputs and outputs, there may be numerous routes from a source node to a destination node. Figure 5.7 shows a number of possible routes between Node A and Node G: A–B–G, A–D–G, A–B–E–G, and A–B–E–D–G, to name only a few.

We can consider our subnet a graph, consisting of nodes (computer nodes) and edges (the communication links between the computer nodes.) If a "weight" or "cost" is assigned to the edge between each pair of nodes, our network graph becomes a **weighted network graph** and takes on a new meaning. Each weight could be interpreted as the dollar cost of using the communication link between the two nodes. The weight could also be interpreted as a time-delay cost associated with transmitting data on that link between the source and destination nodes. There are many algorithms, both actual and theoretical, for selecting a route through a network. While most algorithms strive for an optimal route through a network, some use criteria other than optimality. Below we examine several routing algorithms in no particular order.

Dijkstra's Least Cost Algorithm

One way to select a route through a network is to choose the least expensive path; this can be determined by adding the costs of the path links. The one route with the smallest sum would be the least cost route. Modifying Figure 5.7 to include an arbitrary set of link costs gives us Figure 5.8.

Referring to Figure 5.8, we see that path A–B–G has a cost of 9 (2 + 7), A–D–G has a cost of 10 (5 + 5), and path A–B–E–G has a cost of 8 (2 + 4 + 2). To ensure that we find

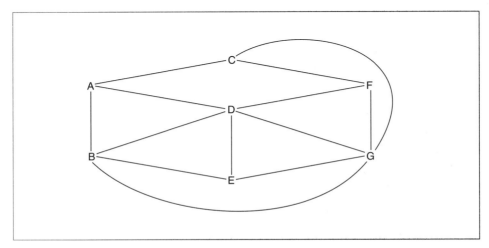

FIGURE 5.7   Sample network with seven nodes in the subnet

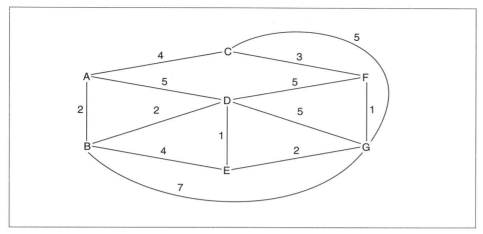

FIGURE 5.8   Network subnet with costs associated with each link

the minimal cost route, we need to use a procedure that will calculate the cost of every possible route starting from a given node. Dijkstra created his **forward search least cost algorithm** to solve this problem. Rather than listing the algorithm in a stepwise form, let's walk through an example solution introducing the algorithm as we proceed.

Using Figure 5.8, we will find the least cost routes from node A to all other nodes. To begin, create a table with a column for each node in the network except the starting node A (Initial Table).

Initial Table

VISITed	B	C	D	E	F	G
	-	-	-	-	-	-

Path:

Next, create a list titled VISITed and initialize the list to null.

VISITed = (null)

Select the starting node A, and VISIT it. Add the starting node to the VISITed list.

VISITed = A

Locate each *immediate* neighbor of node A (only one link away from node A) that has not yet been VISITed (is not in the VISITed list), calculate the cost to travel from node A to each of these neighbors, and enter these values in the table. For example, node B is one link away from node A, it has not yet been VISITed, and traveling from A to B costs 2 units. The route to node B coming directly from A is the least cost path from A to B (it's also the only one so far), so enter 2 in the table along with the path A–B (Update 1).

Update 1

VISITed	B	C	D	E	F	G
A	2	-	-	-	-	-
Path:	A–B					

You can also go from A to C in one link with a cost of 4, and from A to D with a cost of 5 (Update 2). (C and D have not yet been VISITed.)

Update 2

VISITed	B	C	D	E	F	G
A	2	4	5	-	-	-
Path:	A–B	A–C	A–D			

A has no more immediate neighbor nodes that have not been VISITed, so we need to select the next node to VISIT. Let's choose node B, since it has the least cost in our table thus far. Locate the immediate neighbors of node B that have not yet been VISITed (node A has been VISITed) and determine the cost of traveling from node A to each immediate neighbor of B via node B. The immediate neighbors of node B that have not yet been VISITed are D, E, and G. The cost of going from node A to node D via node B is 4 (the link from A to B costs 2, and the link from B to D costs 2). Since this cost is less than the cost of going directly from A to D, replace the value 5 in the table with the new value 4 (Update 3).

Update 3

VISITed	B	C	D	E	F	G
A	2	4	5	-	-	-
A B	2	4	4	-	-	-
Path:	A–B	A–C	A–B–D			

The cost of going from A to E via B is 6 (2 + 4), and the cost of going from A to G via B is 9 (2 + 7). Since there are no entries in the E or G columns, enter the values 6 and 9, respectively (Update 4).

Update 4

VISITed	B	C	D	E	F	G
A	2	4	5	-	-	-
A B	2	4	4	6	-	9
Path:	A–B	A–C	A–B–D	A–B–E		A–B–G

Let's VISIT node C next. The immediate neighbors of C that have not yet been VIS-ITed are F and G. The cost of going from A to F via C is 7 (4 + 3). Since there is no entry in the F column yet, enter the value 7 (Update 5).

Update 5

VISITed	B	C	D	E	F	G
A	2	4	5	-	-	-
A B	2	4	4	6	-	9
A B C	2	4	4	6	7	9
Path:	A–B	A–C	A–B–D	A–B–E	A–C–F	A–B–G

The cost of traveling from node A to G via C costs 9 (4 + 5). As the current value is already 9, there is no need to update the table.

Let's VISIT node D next. The immediate neighbors of D not yet VISITed are E, F, and G. The cost of going from A to E via D (via B) is 5 (4 + 1). Because this value is less than the current cost of going from A to E, you need to update the table (Update 6). The cost of going from A to F via D is 10. The cost of going from A to G via D is 10. As these are more expensive than previously calculated routes, we do not enter these values in the table.

Update 6

VISITed	B	C	D	E	F	G
A	2	4	5	-	-	-
A B	2	4	4	6	-	9
A B C	2	4	4	6	7	9
A B C D	2	4	4	5	7	9
Path:	A–B	A–C	A–B–D	A–B–D–E	A–C–F	A–B–G

The next node to VISIT is E. The immediate neighbor of E not yet VISITed is G. The cost of traveling from node A to node G via node E (via D via B) is 8. This is a smaller value so enter it in the G column (Update 7).

Update 7

VISITed	B	C	D	E	F	G
A	2	4	5	-	-	-
A B	2	4	4	6	-	9
A B C	2	4	4	6	7	9
A B C D	2	4	4	5	7	9
A B C D E	2	4	4	5	7	7
Path:	A–B	A–C	A–B–D	A–B–D–E	A–C–F	A–B–D–E–G

The next node to VISIT is F. The only immediate neighbor of F not yet VISITed is G. The cost of traveling from node A to node G via F (via C) is 8. This is no better than the current value so it is not entered in the table.

The final node to VISIT is G, but since there are no immediate neighbors of G that have not yet been VISITed, we have finished our route cost calculation. The final table is shown in Update 8.

Update 8

VISITed	B	C	D	E	F	G
A	2	4	5	-	-	-
A B	2	4	4	6	-	9
A B C	2	4	4	6	7	9
A B C D	2	4	4	5	7	9
A B C D E	2	4	4	5	7	7
A B C D E F G	2	4	4	5	7	7
Path:	A–B	A–C	A–B–D	A–B–D–E	A–C–F	A–B–D–E–G

Now that the table is complete, you can easily look up the least expensive path from node A to any other node. A data packet originating from node A and destined for node *x* would simply consult column *x* of the table to identify the least cost route. If we wish to find the least-cost path from, say node C, to any other node, we could repeat the least cost algorithm with node C as the starting position to generate the new table.

Flooding

The next routing technique seems quite simple compared to the previous least cost routing algorithm. In **flooding**, each node takes the incoming packet and retransmits it onto every outgoing link. For example, assume a packet originates at node A (Figure 5.9). Node A simply transmits a copy of the packet on every one of its outgoing links. Thus, a copy of the packet is sent to B, C, and D. When the packet arrives at B, node B transmits a copy of the packet to A, D, E, and G. Node C will likewise transmit a copy of the packet to A, F, and G. Node D will also transmit a copy of the packet to A, B, E, F, and G (Figure 5.10). Soon the network will be flooded with copies of the original data packet.

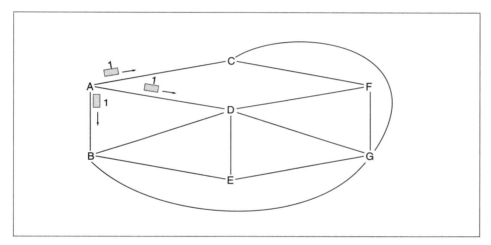

FIGURE 5.9   Network subnet with flooding starting from Node A

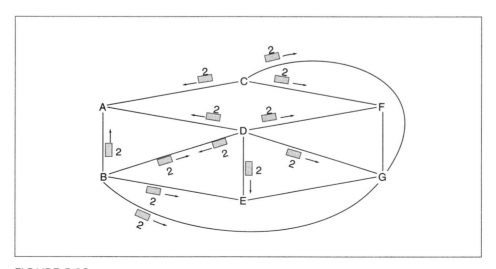

FIGURE 5.10   Flooding has continued to nodes B, C, and D

In order to limit the massive number of copied packets that can result from flooding, a couple of common sense rules must be established. First, a node need not send a copy of the packet back on the link on which the packet arrived. Thus, when node A sends a copy of the packet to C, C does not need to send a copy back to A. Second, a network limit can be placed on how many times any packet is copied. Each time a packet is copied, a counter associated with the packet is incremented. When this counter reaches the network **hop limit**, this particular packet will not be copied anymore. This counter is called a **hop count**. For example, when node A first sends copies to B, C, and D, each of the three copies has a hop count of 1. When the packet arrives at node C, copies with hop counts of 2 will be sent to F and G. When the copy arrives at node F, two copies with hop counts of 3 will be transmitted to D and G. If the network hop limit is set at 3, the packets that arrive at D and G with hop counts of 3 will go no further.

While this may seem like a strange way to route a packet through a network, flooding does have its merits. If a copy of a packet must be sent to a particular node, flooding will get it there, assuming of course there is at least one active link to the particular node and the network hop limit is not set too low. Flooding is also advantageous for getting a copy to all nodes, such as sending emergency information or network initialization information. The major disadvantage of flooding, of course, is the large number of copied packets it distributes throughout the network.

Centralized Routing

The concept of **centralized routing** is not so much an algorithm for routing data packets through a network as it is a technique of providing routing information. Using a previously calculated method such as a least cost algorithm, centralized routing dictates that the routing information generated from the least cost algorithm is stored at a central location within the network. Any node wishing to transmit a packet to another node would consult this centralized site and inquire as to the best route by which to transmit the data. The central site might maintain a table such as the one shown in Figure 5.11. This table was created by performing the least-cost algorithm seven times using each node in the network as the starting node.

The primary advantage of using centralized routing is that all routing information is kept at one site. In this way, there is only one routing table, so you minimize routing table storage and eliminate the possibility of multiple routing tables with conflicting information. A major disadvantage of centralized routing is that you "put all your eggs in one basket." If the node holding the one and only routing table crashes, there will be no routing information for the entire network. Also, if all other nodes must consult the one node holding the routing table, network congestion or a bottleneck may be created at the routing table's node.

Distributed Routing

While centralized routing maintains a routing table at one central location in the network, distributed routing allows each node to maintain its own routing table. When a data packet enters the network at node x, that node consults its own routing table to determine

		Destination node						
		A	B	C	D	E	F	G
	A	-	B	C	B	B	C	B
Origination node	B	A	-	A	D	D	D	D
	C	A	A	-	A	F	F	F
	D	B	B	F	-	E	E	E
	E	D	D	G	D	-	G	G
	F	C	G	C	G	G	-	G
	G	E	E	F	E	E	F	-

FIGURE 5.11   Routing table kept at a centralized network site

	Next node						
	A	B	C	D	E	F	G
Start from node C:	A	A	-	A	F	F	F

FIGURE 5.12   Local routing table for node C

the next node to receive the packet. Since each node needs only the routing information for its own locale, the centralized routing table of Figure 5.11 can be divided by rows and each row assigned to the appropriate node. For example, the routing table for node C from the above examples would be the table shown in Figure 5.12. If a data packet arrives at node C and is destined for node G, the packet should be sent to node F next.

A primary advantage of distributed routing is that no one node is responsible for maintaining a routing table. Thus, if any node crashes, the malfunction will probably not disable the entire network. Second, a node will not need to send a request to a central routing table, since each node has its own table. By eliminating the request packets transmitted to the node holding the routing table, and eliminating the result packets leaving the routing table node, much data traffic is eliminated.

One of the primary disadvantages of using distributed routing information is the problems that arise if the routing tables need to be updated. When all the routing information is in one place, it is simpler to update a single table. With the routing information scattered throughout the network, getting the appropriate routing information to each node is more difficult. Also, with routing information stored at multiple locations, there is always the possibility that one or more routing tables contain old or incorrect information at any given point in time.

Isolated Routing

The previous two routing techniques, centralized and distributed, use routing tables that are created from information from some or all of the other nodes in the network. In the centralized scheme, all nodes could transmit their local routing costs to the central site, which would assemble the centralized routing table. With the distributed scheme, neighbor nodes could share routing information, and then each node could use this information to create its own routing table. What if there is no sharing of routing information between nodes and each node is completely on its own to create its own routing table? This would be an example of **isolated routing**.

With isolated routing, each node must use only local information to create its own table. One technique a node can use to gather information is to observe the status information of incoming packets. If all nodes maintain a clock system and each packet is time-stamped as it leaves a node, when a packet arrives at the receiving node, the receiver can compare the time stamp of the packet with the current time and calculate the time the packet spent in transit. The receiving node can then assume that the time spent coming in on that particular link would be similar to the time a packet would spend going out that same link. The advantage of isolated routing over centralized and distributed routing is that in this method, routing information does not have to be shared and transmitted to any other nodes, thus creating more traffic for the network. Unfortunately, it is also possible to gather the wrong or insufficient data when you are using isolated routing.

Adaptive Routing Versus Fixed Routing

Assuming that some form of least cost routing is used in conjunction with maintaining either centralized or distributed routing tables, the next question that arises is whether the routing tables should adapt to network changes (**adaptive routing**) or stay fixed the entire time (**static routing**). Clearly, a dynamic system such as adaptive routing would react to network fluctuations such as congestion and node/link failure. As a problem occurs in the network, the appropriate information could be transmitted to the routing tables, and new routes that avoid the area of congestion could be created. But how often is the information shared and the routing table(s) updated? And how much additional traffic will be generated by these continuous messages about routing information? Another problem that may occur is for a network to react too quickly to a congestion problem, rerouting all traffic onto a different path and creating a congestion problem in a different area. Maintaining static routing tables is definitely simpler but does not allow quick reaction (and sometimes no reaction) to network problems. As is almost always the case with any option, there are advantages and disadvantages.

NETWORK CONGESTION

When a network or a part of a network becomes so saturated with data packets that packet transfer is noticeably impeded, then **network congestion** has occurred. Congestion may be a short-term problem, such as a temporary line or node failure, or it may be a long-term problem, such as inadequate planning for future traffic needs or poorly created routing tables. As with so many things in life, the network is only as strong as its weakest link. If network designers could properly plan for the future, network congestion might exist only rarely. But the computer industry, like most forms of communication, is filled with examples of failure to plan adequately for the future. We can remember when:

- microcomputers came with a whopping 64 K of memory
- mainframes could address no more than 1–2 megabytes of memory
- only research institutions, select universities, and the government used the wide area network Arpanet (now Internet)
- having an automobile telephone was a luxury reserved for the rich and famous
- telephone area codes had a 0 or 1 as the second digit
- cable television coming into a home had at most 20 channels.

More examples could be listed, but the point is obvious: It is difficult to plan adequately for the future. Computer networks are going to experience congestion, so we may as well consider effective congestion avoidance and congestion handling techniques.

A node failure in the network subnet can be a difficult problem. The network could always require each node site to maintain a backup machine in the event of nodal failure, but this is an expensive solution and one many network sites will reject. As a preventive measure, the network should provide alternate routes so that failed nodes or links could be avoided and routed around. This, of course, implies that routing tables are adaptive and can change to network problems. Many of the more popular networks provide at least two paths between any pair of nodes. In the event of a network problem, routing tables are updated and the problem area is avoided.

Another possible reason for network congestion is insufficient buffer space at a node in the subnet. It is not uncommon to have hundreds to thousands of packets arrive at a network node each hour. If the node cannot process the packets quickly enough, incoming packets will begin to accumulate in a buffer space. When packets sit in a buffer for an appreciable amount of time, network throughput begins to suffer. If adaptive routing is employed, this congestion can be recognized and updated routing tables can be sent to the appropriate nodes (or to a central routing facility). Changing routing tables to reflect congestion, however, might be only a temporary fix. A more permanent solution would be to increase the speed of the node processor responsible for processing the incoming data packets.

Suppose the buffer space is completely filled and a node can accept no further packets? Packets that arrive after the buffer space is full are usually discarded. While this hardly seems fair, it does solve the problem momentarily. A more reasonable solution has already been discussed in the previous chapter on data link protocols. Flow control allows two adjacent nodes to control the amount of traffic passed between them. When the buffer space of a node becomes filled, the node could inform the sending node to stop transmission until later notice.

An alternate solution to controlling the flow of packets between two nodes is **buffer preallocation**. Before one node sends a series of n packets to another node, the first node inquires in advance whether the second node has enough buffer space for the n packets.

If the second node does have enough buffer space, the second node sets aside the n buffers and informs the first node to begin transmission. While this scheme generally works, it does introduce extra message passing, additional delays, and possible wasted buffer space if all n packets are not sent. Of course, the alternative of discarding packets because of insufficient buffer space is worse, so some discomfort is not unreasonable.

Flow control and preallocation of buffers help control congestion between nodes, but what can be done to eliminate congestion between a station and a node? Similar to the concept of flow control between two nodes is the concept of **choke packets**. As we saw in Figure 5.3, one node may have multiple stations connected to it. The node continuously monitors the amount of traffic arriving from each station. If one station begins to transfer data packets at a higher rate than the node can accommodate, the node will use a choke packet to inform the station to slow down.

Other techniques have been proposed for limiting the number of packets traveling throughout the network. If the total number of outstanding packets in transit in a network can be limited, it should be possible to reduce or eliminate congestion. Thus, some type of **permit** system is created whereby a station may not enter a packet into the network unless it possesses a permit. While this sounds attractive, it can bring problems:

- What is the process for determining who gets permits and how many?
- How do you keep the permits equally distributed?
- What do you do with permits you don't need?
- What happens if one or more permits are lost?

All these problems dull the attractiveness of such a plan.

Finally, limiting the number of packets/permits in a network does not necessarily eliminate congestion. What happens if all the permits end up at one node? There may still be congestion in the area of the one node, while there is little or no traffic anywhere else in the network.

CASE STUDY

As we have learned in this chapter about networking, so have the staff in our case study. The computing center at XYZ University was just introduced to the basic concepts of networks and immediately, the first question that came to everyone's mind was, "Should we be considering some type of network connection for our campus?" This question was followed with, "But what type of network?" As we will soon discover, a local area network is a possible solution for interconnecting computer workstations that are in close proximity to one another. Perhaps the computing center could replace some or all of the terminals on campus with microcomputer workstations, then interconnect them with a local area network? Replacing each terminal with a microcomputer would be fairly expensive, but maybe all of them would not have to be replaced. For example, staff could start with the faculty terminals and replace those with microcomputers, then start replacing the terminals used for computer support. The advantage of these replacements is that a microcomputer workstation is more powerful than a terminal because it can execute its own software independent of the mainframe computer.

A second application area for the computing center to consider is replacing the leased phone line connection to State University with some type of network connection. The leased phone line is fairly expensive and is not used continuously. If there were some type of network that the two universities could connect to, it would represent a less expensive solution.

A third possible network application is connecting the university's computer to the nationwide network called Internet. Somebody on the XYZ University computing center staff read that the Internet provides access to thousands of other computer sites, both national and international. Through the Internet, XYZ University would have access to numerous online databases, library card catalogs, news services, and colleagues throughout the world. With the anticipation that the United States will creat an "information superhighway" some day, the XYZ computing staff does not want to be left out in the cold.

Clearly, the computing center has much to consider. Before making any decisions, however, the staff have decided they need more information and will read on.

SUMMARY

The third layer of the OSI model is the network layer and is responsible for transmitting data packets through a computer network in a timely fashion. The network layer must also ensure that the network does not become congested because of network failures or too many data packets.

To be meaningful, the terms that describe the various characteristics of computer networks must be understood in relation to the concepts of connectionless and connection-oriented networks, the relationship between a network station and a network node, and the differences between a datagram network and a virtual circuit network.

Selecting the optimal route for the transfer of a data packet through a network is a common service of many networks. Many routing algorithms exist, both actual and theoretical. One possible way to select a route through a network is to choose a path whose path link costs have the smallest total value. This technique is based on Dijkstra's least cost algorithm and is a common method for determining the optimal route.

Flooding is a routing technique that requires each node to take the incoming packet and retransmit it onto every outgoing link. If a copy of a packet must be sent to a particular node, flooding will get it there. Unfortunately, flooding creates a very large number of copied packets that are distributed throughout the network.

The concept of centralized routing is not so much an algorithm for routing data packets through a network as it is a technique of providing routing information. Centralized routing dictates that the routing information generated by a method such as the least cost algorithm be stored at a central location within the network.

Distributed routing allows each node to maintain its own routing table. Isolated routing, fixed routing, and adaptive routing allow a network to establish routing tables.

When a network or a part of a network becomes so saturated with data packets that packet transfer is noticeably impeded, then network congestion has occurred. Congestion may be the result of network node failure, network link failure, or insufficient nodal buffer space. Remedies for network congestion include discarding excess nodes, preallocation of nodal buffers, flow control, permit systems, and choke packets.

EXERCISES

1. State the primary difference between a connection-oriented network and a connectionless network.
2. Which type of network requires more elaborate software: connection-oriented or connectionless? Explain.
3. Create a scenario, similar to the telephone call/sending a letter scenarios (Figures 5.4 and 5.5) that demonstrates the differences between connection-oriented networks and connectionless networks.
4. Explain the difference between a network station and a network node.
5. What is the difference between a datagram subnet and a virtual circuit subnet?
6. List the steps involved in creating, using, and terminating a virtual circuit.
7. Even though the transport layer is not discussed in detail until a later chapter, state the relationship between the transport layer and the network layer as you understand it at this point.
8. How is a weighted network graph similar to many actual networks, such as the telephone network?
9. Given the graph for Exercise 9, find the least cost route from node A to all other nodes using Dijkstra's least cost algorithm.

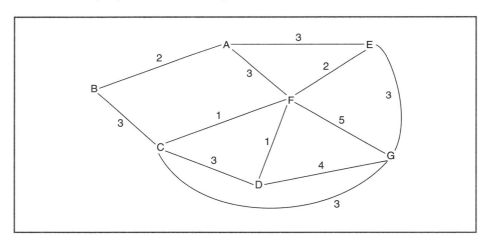

10. Using flooding and the graph for Exercise 9, how many packets will be created if there is a network hop limit of three?
11. Using the least cost solution from Exercise 9, create the centralized routing table similar to one shown in Figure 5.11.
12. What are the advantages of centralized routing over distributed routing?
13. What are the advantages of distributed routing over centralized routing?
14. It was stated that isolated routing may gather the wrong or insufficient data. Explain how this may happen.
15. List five computer-related examples of failure to plan adequately for the future (different from those listed in the text).
16. The text suggested four possible reasons for network congestion. List them. Can you offer any other reasons?

CHAPTER 6

LOCAL AREA NETWORKS

INTRODUCTION

A local area network (LAN), by definition, is a communication network that interconnects a variety of *data communicating devices* within a *small area* and transfers data at *high transfer rates* with *very low error rates*. There are several points in the above definition that merit further discussion. The phrase *data communicating devices* can include computers, peripherals (such as disk drives, printers, and modems), and other data transferring devices (such as data collecting equipment, and motor and speed controls).

Within a *small area* usually implies that LANs can be as small as a couple of meters in length or extend up to 50 kilometers. Networks extending past 50 kilometers often fall into the category of Metropolitan Area Network. Most often a LAN is found within the confines of a single room or building.

High transfer rates includes the range of 0.1 Mbps up to 100 Mbps, with some special application LANs extending beyond 100 Mbps. While LANs have typically operated in the 10 Mbps range, the growth of fiber optic transmission has pushed the transfer rates well past 10 Mpbs. *Very low error rates* implies that LANs possess error rates no greater than 1 lost bit in 10^8 to 10^{11} bits transmitted. While all these values are typically found in average LANs, you can always find an example that deviates from this norm. However, these values describe the basic characteristics common to most local area networks.

Perhaps the strongest advantage of a local area network is its ability to share hardware and software resources. For example, a special version of a popular word processor program can be purchased and installed on a LAN, and all users of the LAN can then use that word processor. Similarly, a high-quality printer can be installed on the LAN with all members sharing access to this expensive peripheral.

Other advantages of local area networks include but are not limited to the following:

- An individual workstation can survive a network failure, but only if the workstation does not rely on software or hardware found elsewhere on the network.
- Component or system evolution, independent of each other, is often possible.
- Under some conditions, equipment from different vendors can be mixed on the same network.
- Because of the high transfer rates, documents can be transmitted rapidly across the network.
- Most local area networks provide some means of interconnection to wide area networks and mainframe computers.
- Since local area networks can be purchased outright, the entire network and all workstations/devices can be privately owned.

Unfortunately, there are also disadvantages to local area networks:

- Local area networks, their operating systems, and the software that runs on the network can be expensive. One common, expensive device found on most LANs is the file server, a large, high-speed disk storage device that houses shared software. Hidden costs also include cables, taps, any amplifiers or repeaters, installation and training costs, and maintenance costs.
- Despite the assumption that a local area network can support any type of hardware or software, this is not always true. Some software products are designed to operate only on a single computer workstation, not on a network that can be shared by many workstations. In order for a software package to operate on a local area network, you may have to buy a special network version or network license. Most often, these versions and licenses are more costly than single workstation versions.
- Management and control of the local area network requires many hours of dedication and service. A manager of a LAN should not assume that the network can be maintained with just a few hours of attention per week.
- Just because two different database packages can operate on the same local area network does not mean that their data are interchangeable. While the two packages may be able to communicate with each other, their data formats remain unique.
- A network is only as strong as its weakest link. A LAN, for example, may suffer terribly if the file server cannot adequately serve all the requests from users of the network.

When you understand these advantages and disadvantages, you see that the decision to incorporate a local area network into an existing environment is a decision that you should not take lightly.

Functions of a Local Area Network

To help you decide whether a local area network is desirable for certain situations, you need to know what they do. Next we examine the typical functions and applications of a LAN:

- **File serving**—A local area network, with the aid of a file server, can act as a large storage device for shareable programs and data sets. If the network offers a word processing program, the software for that program will reside on the file server disk until requested by a user workstation. At the time of invocation, all or part of the software will be transferred to the user workstation for execution.

- **Print serving**—It is also very common for the network to have access to a high-quality printer. Any workstation may then route its output to this printer. Network software queues print jobs if necessary and print any appropriate banners.
- **Electronic mail**—Many LANs can provide the service of sending and receiving electronic mail.
- **Process control and monitoring**—In manufacturing and industrial environments, LANs are often used to monitor events and report their occurrence or invoke a response to an event.
- **Remote links**—Most LANs have the capability to interface with other LANs, wide area networks, and mini- and mainframe computers.
- **Teleconferencing**—Certain LANs can support teleconferencing, including video transmission.
- **Distributed processing**—Depending on the type of network and the choice of network operating system, a LAN may support distributed processing, in which a task is subdivided and sent to remote workstations on the LAN for execution. The results of these remote executions are then returned to the originating workstation for dissemination or further processing.

These functions show that the LAN can be an effective tool in many application areas. One of the most common application areas is the electronic office. A local area network in an electronic office can provide the capability of word processing over a wide variety of platforms and at a large number of workstations. Completed documents can be routed to high-quality printers, which can produce letterheads, graphically designed newsletters, and formal documents.

Office workstations can also be connected to an electronic mail service. The E-mail system can deliver internal mail as well as interface with external mail services. Scheduling programs can allow individuals and groups to synchronize meeting times and appointments.

A second common application area for a local area network is manufacturing. Modern assembly lines operate exclusively under the control of local area networks. As products move down the assembly line, sensors control position, robots perform exacting or mundane operations, and product subassemblies are inventoried and ordered. The modern automobile assembly line is a technological tour de force incorporating numerous local area networks and mainframe computers.

A third common application area for a local area network is education. Formal and informal classroom settings are often designed around local area networks. Tutorials and lessons can be presented using high-quality graphics and sound. Multiple workstations can provide student-paced instruction as the teacher monitors and records the progress of each student at every workstation.

BASIC TOPOLOGIES

Local area networks are often interconnected into one of four basic configurations, or **topologies**. The choice of topology is often dictated by the environment in which the

LAN is to be placed or by the software (medium access control protocol) chosen. The **medium access control protocol** determines how an individual workstation places its data onto the network for transmission and is covered in more detail in the next section.

BUS/TREE TOPOLOGY

The bus topology was the initial topology when LANs became commercially available in the late 1970s and is still the most widely used. As is shown in Figure 6.1, the bus is simply a linear cable that multiple devices or workstations tap into.

Since all stations communicate through this single cable, steps must be taken to ensure that two or more stations do not attempt to transmit a signal at the same frequency at the same moment. This control is achieved by a medium access control protocol. But even something as simple as a linear bus is complicated by the fact that two different technologies exist for transmitting a signal on the bus. The first is **baseband technology**. With this technique, a single digital signal (such as Manchester encoding) is transmitted over the bus. This digital signal uses the entire spectrum of the cable, allowing only one digital signal at a time on the cable. All workstations must be aware that someone else is transmitting so they do not attempt to transmit and destroy the signal of the first transmitter. Allowing only one workstation access to the medium at one time is the responsibility of the medium access control protocol.

There are other characteristics of baseband technology worth noting:

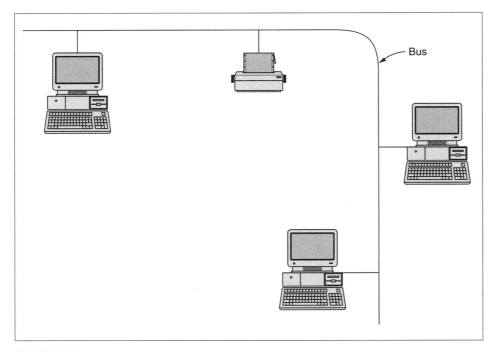

FIGURE 6.1   Simple diagram of a bus topology for a LAN

- Baseband transmission is **bidirectional**; when the signal is transmitted from a given workstation, the signal propagates in both directions on the cable away from the source (Figure 6.2).
- Coaxial cable was originally the most popular choice for the medium, but due to advances in technology (and cost), twisted pair cable (shielded and unshielded) is gaining in popularity.
- The cable/network may extend for hundreds of meters in length. This distance is highly dependent on factors such as signaling rates and noise in the environment.
- It is typical to see baseband LANs limited to a maximum of 100 attached workstations/devices, with most networks much smaller.
- The average baseband LAN transmits data at a rate rarely exceeding 10 Mbps. However, technological increases such as fiber-optic transmission are pushing the data transmission rates into the hundreds of Mbps.
- Baseband LANs are relatively easy to install and maintain, requiring minimal physical maintenance and service.

The second type of technology used on the bus local area network is **broadband technology**. This method uses analog signaling in the form of frequency division multiplexing to divide the available medium into multiple channels. Each channel is capable of carrying a single conversation between two workstations. If the medium can be divided into multiple channels, broadband signaling allows multiple concurrent conversations.

Since a broadband signal is transmitted by frequency division multiplexing, analog frequencies need to be amplified often to maintain an acceptable signal, and since analog amplifiers work in only one direction, broadband transmissions are unidirectional. To establish a half- or full-duplex conversation between sender and receiver, two sets of cables, or their equivalent, are required. Figure 6.3 shows two possible configurations to allow this type of conversation. Figure 6.3a demonstrates two cables, with one cable transmitting a signal in one direction, and the second cable transmitting a signal in the opposite direction. Each workstation would require two taps, one for each cable.

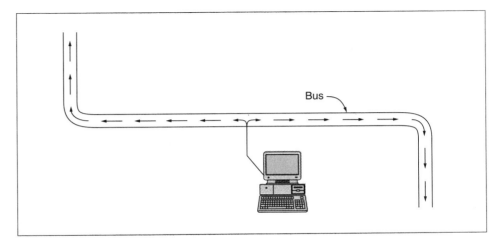

FIGURE 6.2   Bidirectional propagation of a baseband signal

In Figure 6.3b, a single cable is looped around, with each workstation still requiring two taps, one for each direction.

Below are several advantages that broadband has over baseband as well as some further characteristics of broadband technology:

- Broadband signals are relatively easy to amplify, a trait that increases the maximum distance or length of a broadband network.
- Because broadband signals have such a wide bandwidth, from which multiple channels can be created, broadband can support data, video, and radio signal transmissions.
- It is possible to split and join broadband cables and signals to create configurations more complex than a single linear bus. These more complex topologies are termed **trees** (Figure 6.4).
- Broadband technology uses standard, off-the-shelf 75-ohm cable television parts.
- A single broadband network is capable of supporting hundreds to thousands of workstations and/or devices.

There are also disadvantages of broadband technology compared to baseband technology:

- Because of the nature of analog frequencies, broadband networks are more expensive than baseband networks to install and maintain.

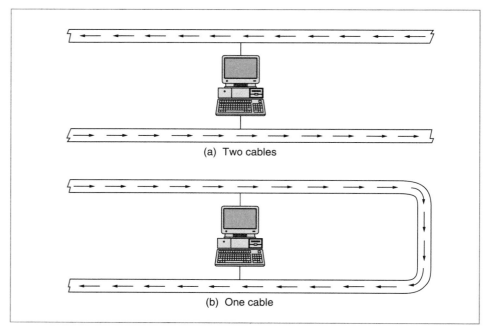

(a) Two cables

(b) One cable

FIGURE 6.3   Interconnection of broadband workstations to allow half- or full-duplex connections

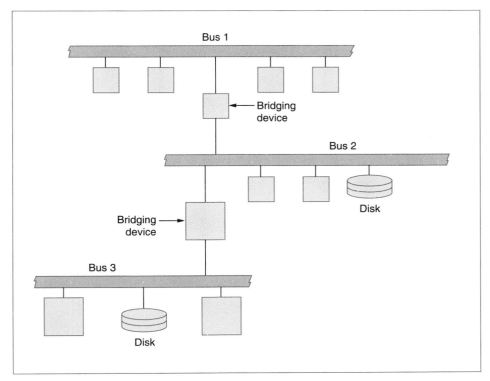

FIGURE 6.4   Simple example of broadband tree topology

- A specialist is needed to "tune" the network periodically to ensure that the analog frequencies of the individual channels remain as originally designed.
- Because broadband transmissions are unidirectional, the average propagation delay is twice that of baseband technology.

Star Topology

Another possible configuration for a LAN is the star network. This topology is a simple design, as shown in Figure 6.5. In it, all workstations and devices connect to a central hub by a unique point-to-point line. This central hub can also act as the local data storage (file server), but often the file server is a separate unit. Twisted pair cabling has become the preferred medium, and installation and maintenance are relatively straightforward. For one workstation to send information to a second workstation, all communications must pass through the central hub. The central hub examines the address of the intended receiver and forwards the message to that workstation. Thus, all connections between workstations are composed of two direct links: the link from the sending workstation to the hub, and the link from the hub to the receiving station.

The major disadvantage of the star design is the potential bottleneck that may result because all transmitted messages must travel through the central hub.

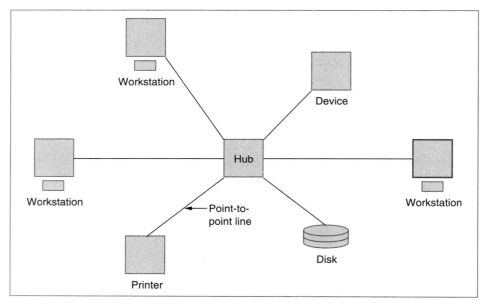

FIGURE 6.5   Diagram of a star topology local area network

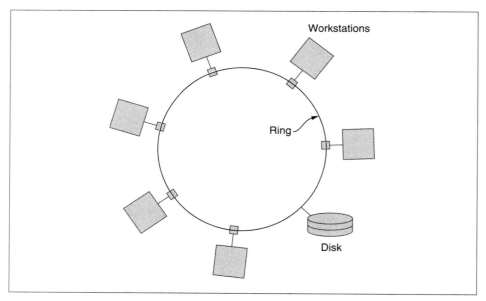

FIGURE 6.6   Ring topology as seen from a distance

Ring Topology

The ring topology, when viewed from a distance, is a circular connection of workstations, as Figure 6.6 demonstrates. On closer examination, however, we see that the ring is actually composed of segments of one-way point-to-point cables strung between pairs of repeaters (Figure 6.7).

Each workstation/device is attached to a repeater and receives data from the network or places data onto the network through its repeater. A basic repeater is an electronic device that accepts a digital signal, performs a regeneration of the signal, and then transmits the new signal. Thus, a noisy digital signal is regenerated into a noiseless digital signal (Figure 6.8). Most repeaters operate very quickly and can regenerate a signal in

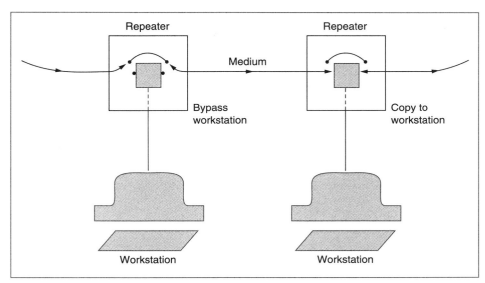

FIGURE 6.7   Close-up view of ring repeaters and workstations

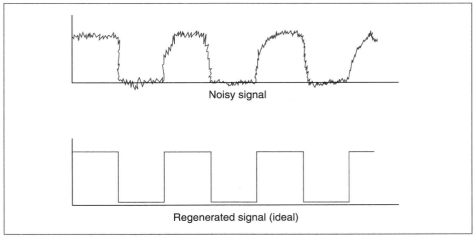

FIGURE 6.8   Noisy digital signal regenerated by a repeater

the time period between two successively received bits. Thus, we say that the repeater introduces a transmission delay equivalent to the "duration of one transmitted bit."

A ring repeater can perform three other functions besides regenerating a digital signal. First, it can copy any data presently on the ring and send the copy to its attached workstation. This function allows the workstation to accept or input data from the ring. Second, the repeater can insert or place new data from the workstation onto the ring, allowing the workstation to transmit data to other workstations. Third, the ring repeater can completely ignore any data traversing the ring, essentially removing the workstation from the ring.

The ring topology has one or two characteristics that make it especially attractive when compared to other local area network topologies:

- Since repeaters are fast (only a one-bit delay), data travels around the ring quickly.
- Under certain topological circumstances, the ring configuration may "fit" better than either the bus or star topologies.
- It is common for a ring configuration to require less cabling than an equivalent star configuration.

As with any technology, there are disadvantages to the ring topology. As data circulates around the ring passing into and out of repeaters, signal timing errors increase. These timing errors result primarily from delay distortion and imperfections in the repeater circuitry. As these timing errors increase, **timing jitter** develops, distorting the original signal.

A second disadvantage of the ring topology is that data circulates on the ring in only one direction. A workstation wishing to transmit data to a neighbor workstation "upstream" must pass the data completely around the circle. Luckily, data transfers very quickly (recall that repeaters cause only a single-bit delay). A break in the cable can present a more serious problem. Since data is transferred in one direction only, a cable break disables the entire network. This may not be the case in a star or bus network.

Star-Ring Topology

The **star-ring** topology is a hybrid design that combines the advantages of both the star and ring topologies while avoiding most of their disadvantages. One of the key players in this design is the **ring wiring concentrator** (Figure 6.9). The ring wiring concentrator is a clever device that interconnects typically up to eight workstations.

When no workstations are plugged into the concentrator, all external connections are closed and an internal loop (or ring) is created. When a workstation is connected to the ring wiring concentrator, the loop is automatically extended into the workstation. Carrying this concept further, it is possible to interconnect multiple ring wiring concentrators, effectively creating one large loop or ring.

The star design is created when multiple rings are interconnected by a **bridge**. The bridge, which has a small amount of intelligence, can filter, buffer, and pass data from one ring to the next (Figure 6.10).

The star-ring design is more secure against failure than other LANs because of its ability to isolate segments of faulty hardware by allowing the ring wiring concentrators

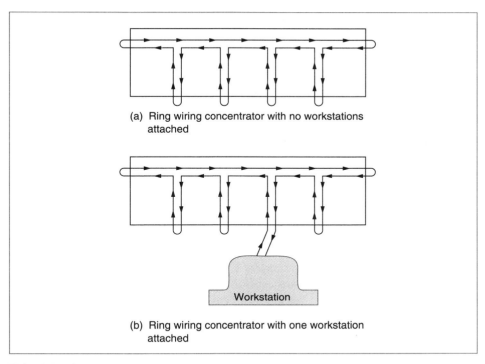

(a) Ring wiring concentrator with no workstations attached

(b) Ring wiring concentrator with one workstation attached

FIGURE 6.9   Ring wiring concentrator

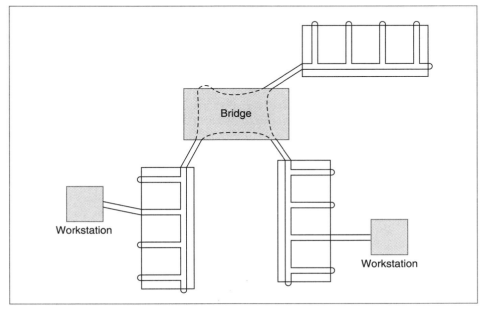

FIGURE 6.10   Star-wired ring topology

to "collapse" the star-ring away from the faulty segment. The price for this ability, however, is the need for costly ring wiring concentrators and bridging devices.

MEDIUM ACCESS CONTROL PROTOCOLS

A medium access control protocol is the software that allows a workstation to place data onto the network. Depending on the network's topology, several types of protocols may be applicable. Since we introduced the bus topology first, here we discuss the two medium access control protocols that are associated with the bus.

Carrier Sense Multiple Access with Collision Detection (CSMA/CD)

The name of this protocol is so long that it almost explains itself. With this access protocol, the workstation "listens" to the medium (senses for a carrier) to learn whether anyone else is transmitting. If no one is, the workstation transmits its data onto the medium. As the data is being transmitted, the workstation continues to listen to the medium. Under normal conditions the workstation should just "hear" its own data being transmitted. If the workstation hears garbage, a collision is assumed. A collision results when two or more workstations listen to the medium at the same moment, hear nothing, and then transmit their data at the same moment. If the signals are transmitted at the same frequencies (always in baseband, and likely in broadband), a collision of signals results.

What are the chances that two workstations will transmit their data at the same moment? The likelihood is not very high, but it happens, particularly when the network is under a heavy load and many workstations are trying to access it simultaneously. Consider the situation in which two workstations are at opposite ends of the bus. A signal propagates from one end of the bus to the other in time n. Since a workstation is not supposed to transmit if the medium is busy, the only way a collision could happen is for one workstation to start a transmission as the signal from another workstation is propagating down the bus. The time during which a collision could happen is the called the **collision window**.

Suppose a workstation wishes to transmit and listens to the medium. What if the medium is busy? The workstation obviously does not transmit, but waits. How long does the workstation wait? Or should we say, what degree of **persistence** does the workstation exhibit? Three different persistence algorithms have been created:

1. **Nonpersistent**—if busy, wait time t then listen again.
2. **1-persistent**—if busy, continue listening until not busy, then send immediately.
3. **p-persistent**—if busy, continue listening. When the medium becomes idle, transmit with probability p (or delay "one time unit" with probability $1 - p$).

The nonpersistent algorithm is as its name implies. If the medium is busy, the workstation tries again later. What if the medium had become free immediately after the workstation listened and had found it busy? That would be too bad; if the workstation had been more persistent, it would have learned sooner that the medium had become free and it would not have wasted time waiting. The 1-persistent algorithm takes this condition into account and listens continuously until the medium is free, then transmits immediately. But what if two workstations following the 1-persistent algorithm are listening and waiting? They will probably try to transmit at the same moment causing a collision. The *p*-persistent algorithm is a compromise of the first two. The workstation continuously listens until the medium becomes free, but then does not transmit immediately. It transmits only with probability *p*. If $p = 0.1$, then nine times out of ten the workstation waits, while one time out of ten it transmits. The odds are in our favor that if two or more workstations are waiting for a free medium, they both will not begin transmitting at the exact moment the medium is idle.

The CSMA/CD access protocol is analogous to humans carrying on a conversation. If no one is talking, someone can speak. If someone is talking, everyone else hears this and waits. If two people start talking at the same time, they both stop immediately and back off a certain amount of time before trying again. If only more people followed these rules!

Token Bus

The token bus medium access control protocol was designed to operate on a bus; but unlike CSMA/CD, it was created to provide a deterministic access to the network. In order for a workstation to transmit data onto the bus, the workstation must possess a software token. Since there is only one token in the entire network, only one workstation may transmit at a time. When the workstation has completed its transmission, the token is passed on to the "neighbor" workstation. Each workstation maintains a list of neighbors, thus creating a defined order of token passing. A neighbor need not be a physical neighbor, only a logical neighbor. Thus, the token bus protocol creates a token passing logical ring on top of a physical bus. The token bus can also create a priority scheme in which higher-priority stations may seize the token from lower-priority stations.

The unique combination of physical bus and logical ring give the token bus protocol several advantages over CSMA/CD:

- Since only the workstation holding the token can transmit, there is no need for the workstation to listen for a collision while transmitting, because collisions should not occur.
- The concept of priorities introduces a flexibility that CSMA/CD does not possess.
- It is possible for a workstation to hold the token for multiple messages rather than relinquishing it after one transmission. This feature could be advantageous when the system is under light load and one workstation has a high volume of data to transmit.
- Since the passing of the token allows each station (eventually) to transmit on the bus and there are no collisions, network performance under heavy loads is much better than with CSMA/CD, and wait-time to transmit is determined rather than by chance.

One major disadvantage of the token bus protocol is the complexity of the software needed to maintain the token. This software has to address important qustions such as what happens if the token gets lost? Who generates the new token? When a new workstation comes online or an existing workstation goes offline, how are the neighbors informed of this change?

Token Ring

The token ring medium access control protocol is functionally quite similar to the token bus. There is one software token that is passed around the ring from workstation to workstation. Only the workstation holding the token may transmit data onto the ring. Unlike the token bus, however, the token ring always passes the token to its next physical neighbor "downstream." Similar to the token bus, the token ring has a priority system for seizing the token from lower-priority workstations.

The token ring protocol has many of the same advantages and disadvantages as the token bus protocol. Maintenance of the token is difficult, requiring complicated software. The priority system can be advantageous but it also requires complex software and may cause lower-priority workstations to wait longer periods of time than they would with no priority system. The ability for each station to transmit is determined by a protocol, however, and throughput under heavy loads is quite good.

IEEE 802 Standards

Beginning in 1980, the IEEE Computer Society Local Network Standards Committee began work on a project to develop *one* IEEE standard for the interconnection of computers and associated peripherals in a local area. Over two decades later, we see instead of one standard a plethora of standards, and more coming every day. Figure 6.11 shows the basic status of IEEE local area network protocols as of approximately 1990.

Note that the IEEE 802 standard encompasses two layers of the OSI model: the physical layer and the data link layer, which for local area networks has been subdivided into the medium access control (MAC) sublayer and logical link control (LLC) sublayer. Since the choice of MAC sublayer protocol depends on the choice of the hardware at the physical layer, there is a close relationship between the MAC sublayer and the physical layer.

Logical Link Control (LLC) sublayer	Defines two Service Access Points, one for sender and one for receiver.			
Medium Access Control (MAC) sublayer	802.3 CSMA/CD	802.4 Token bus	802.5 Token ring	FDDI Fiber optic ring
Physical layer	Baseband coaxial 10 Mbps Twisted pair Unshielded 1, 10 Mbps	Broadband coaxial 1, 5, 10 Mbps Carrierband 1, 5, 10 Mbps	Twisted pair Shielded 1, 4 Mbps	Optical fiber 100 Mbps

FIGURE 6.11   IEEE 802 local area network protocols

The **IEEE 802.3** standard for CSMA/CD comes in several types. The first choice is 50-ohm coaxial cable transmitting a baseband signal using Manchester encoding at 10 Mbps for 500 meters maximum segment length. This is the original design and is called 10Base5. The second choice is 50-ohm baseband (Manchester) for 10 Mbps at 185 meters and is termed 10Base2, or CheaperNet. The third choice of CSMA/CD is unshielded twisted pair cable using baseband (Manchester) at 1 Mbps for 250 meters and is called 1Base5; it is used on the Starlan local area network. The fourth choice is unshielded twisted pair baseband (Manchester) for 10 Mbps for 100 meters and is termed 10BaseT. The final choice of CSMA/CD uses 75-ohm coaxial cable with broadband signaling (differential phase shift keying) at 10 Mbps for 3600 meters and is termed 10Broad36.

When a station transmits a packet of data onto the bus, the medium access control sublayer packet format shown in Figure 6.12 is used. Most of the fields shown in Figure 6.12 are self-explanatory except for the PAD field. The minimum size packet that any station can transmit is 64 bytes long. Packets shorter than 64 bytes are considered packet fragments that resulted from a collision and are automatically discarded. Thus, if a workstation wishes to transmit a packet in which the data field is very short, PAD characters are added to ensure that the overall packet length equals at least 64.

The **IEEE 802.4** standard for a token bus has likewise been subdivided into multiple configurations. A broadband designation has been created that employs 75-ohm coaxial cable using amplitude phase shift keying at 1, 5, or 10 Mbps with no specified maximum length. Single-channel carrierband uses 75-ohm coaxial cable with broadband signaling (frequency shift keying) at 1, 5, and 10 Mbps for a maximum segment length of 7600 meters. The third version of token bus has been defined for optical fiber using amplitude shift keying/Manchester encoding at 5, 10, and 20 Mbps for no specified length.

The medium access control sublayer packet format for IEEE 802.4 is shown in Figure 6.13. In this figure, PRE is the preamble field that consists of 1 or more bytes of

PREAMBLE	START OF FRAME BYTE	DESTINATION ADDRESS	SOURCE ADDRESS	DATA LENGTH	DATA	PAD	CHECKSUM
7 bytes of 10101010	10101011	2 or 6 bytes	2 or 6 bytes	2 bytes	0–1500 bytes	0–46 bytes	4 bytes

FIGURE 6.12   Packet format for IEEE 802.3 CSMA/CD

PRE	SD	FRAME CONTROL	DESTINATION ADDRESS	SOURCE ADDRESS	DATA	CHECK	END DELIM
>=1 byte	1 byte	1 byte	2 or 6 bytes	2 or 6 bytes	0–8182 bytes	4 bytes	1 byte

FIGURE 6.13   Frame format for IEEE 802.4 Token Bus

synchronization bits. SD is the starting delimiter and consists of the bits VV0VV000, where the V bits are code violation bits, such as sending a Manchester code with no transition in the bit frame. The FRAME CONTROL field contains information concerning the frame, such as, does this frame contain data or is it simply a token? END DELIM is the ending delimiter and has the binary configuration VV1VV11E, where V is a code violation bit and E is an error bit that can be set to 1 by a repeater if a checksum error has been detected.

As we discussed earlier under medium access control protocols, fairly involved software is required to support the network and maintain the token. At the least, one or more workstations on the bus must be able to perform the following tasks:

- Initialize the network at system startup or after network failure
- Add a new node to the network and inform neighbors of addition
- Delete a node from the network and inform neighbors of deletion
- Provide a priority system for the token and nodes on the network
- Provide for recovery from error conditions such as loss of token, token hogging, and duplicate tokens

The **IEEE 802.5** standard for token ring local area networks is simpler than either 802.3 or 802.4. It is presently defined only for shielded twisted pair cable using baseband signaling with Differential Manchester encoding at 1 or 4 Mbps with a maximum of 250 repeaters. The sender of the packet is also the workstation responsible for removing the packet from the ring. Like the 802.4 standard for token bus, the software to support the token ring is complex, especially when the optional priority system is invoked.

The packet format for the token ring is shown in Figure 6.14. The SD field is the starting delimiter and is VV0VV000 in binary where V is a code violation bit. The AC (Access Control) field and the FC (Frame Control) field together provide priority information along with other control information. The ED (Ending Delimiter) is VV1VV1IE in binary, where V is a code violation bit, I is an Intermediate Frame bit (0 means last or only frame of transmission, 1 means more frames follow), and E is an error detected bit that can be set to 1 by a bridge if a checksum error is detected. The final field, the FS (Frame Status) field contains the bit sequence ACRRACRR. The A bit is set to 1 by the receiving station if the address for which the packet is intended is recognized. The C bit is set to 1 if the receiving station also copies the packet (accepts the packet). The R bits are reserved for future use. Since a packet is removed from the ring by the sending workstation, the A and C bits can be set by the receiving worksta-

SD	AC	FC	DEST ADDR	SOURCE ADDR	DATA	CHECKSUM	ED	FS
1 byte	1 byte	1 byte	2 or 4 bytes	2 or 4 bytes	n bytes	4 bytes	1 byte	1 byte

FIGURE 6.14   Packet format for IEEE 802.5 Token Ring protocol

tion to inform the sender quickly if the packet was recognized and if the packet was accepted. Note there are two A bits and two C bits. The second bit of each is simply a redundant copy of the first to provide a small degree of error protection

FDDI (Fiber-Distributed Data Interface)

The **FDDI protocol** resembles a token ring network that has been lifting weights. With a data transmission speed of 100 Mbps, network distances up to 200 km, and a possible interconnection of 1000 stations, an FDDI network is a vastly updated token ring network and is considered a high-speed local network. To achieve these impressive figures, several changes were made to the original token ring design. The first and major change is the use of fiber optic cable for the network medium. Fiber optic cable, as we saw in Chapter 2, allows very high data transfer rates with very low error rates.

The second modification was to create a topology of a ring within a ring (Figure 6.15). The second ring transmits data in the opposite direction of the first ring and acts as a backup to the first. A workstation is either a class A station and connects to both rings, or is a class B station and connects to only one ring.

The next improvement was to use the newer 4B/5B encoding over Differential Manchester encoding. You recall the advantage of 4B/5B of not having a signal transition in the middle of each bit frame; thus its baud rate is not twice the bps. At the higher transmission rate of 100 Mbps, a baud rate double the bits per second is significant.

Finally, the token passing protocol was modified. In standard token ring, a workstation seizes the token, transmits its data, removes its data from the ring, and passes the token on to the next workstation. FDDI seizes the token, transmits its data, which may

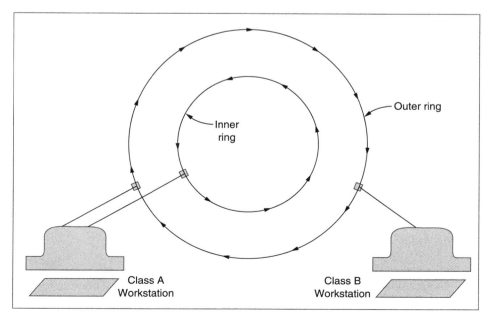

FIGURE 6.15   FDDI dual ring topology

consist of multiple packets, and appends the token to the end of the last data packet. As the data packet travels to the next neighbor, the neighbor workstation can strip the token off the end of the neighbor's packet and transmit its own packet. The reasoning is that data transmission is so fast and distances possibly so long that too much time would be wasted using the older method.

Shielded Twisted Pair FDDI

Three variations of the 100 Mbps fiber optic FDDI standard have been created in the last few years. The first variation, known as the Green Book, replaces the fiber-optic medium of FDDI with shielded twisted pair cable, yet maintains the 100 Mbps data transfer speed of the original. The second variation is the IBM-supported Shielded twisted pair Distributed Data Interface (SDDI), while the third variation is Copper Distributed Data Interface (CDDI). Both SDDI and CDDI have formats similar to FDDI, maintain the 100 Mbps data transfer speed, but use either shielded twisted pair or unshielded twisted pair cable in place of fiber-optic cable.

The ability to transmit FDDI over shielded and unshielded twisted pair cable offers a number of advantages. The most obvious advantage is the lower costs of shielded twisted pair cable over fiber-optic cable. A second advantage is the ability of an existing shielded twisted pair network, such as 10BaseT, to be upgraded to FDDI by replacing the supporting hardware.

A major disadvantage of the nonfiber FDDIs is the difficulty of transmitting data at such a high rate over twisted pair cable. Since twisted pair can act like an antenna, emitting various frequencies while absorbing others, nonfiber FDDIs use unique transmission methods that spread out high levels of energy normally concentrated at particular frequencies.

EXAMPLE LOCAL AREA NETWORKS

Ethernet

The most popular form of local area network on the market today is Ethernet. It was originally developed by Xerox Corporation in the mid-1970s and later modified by Digital, Intel, and Xerox working together. The IEEE 802.3 CSMA/CD protocol was designed to operate as Ethernet, and many consider the two protocols to be equivalent. Note, however, that the original Ethernet is slightly different from the IEEE 802.3 standard. In the original Ethernet, all source and destination addresses were six octets, while 802.3 allows either two- or six-octet addresses. Fortunately, a six-octet address under 802.3 is the norm. Second, the original Ethernet had a two-octet TYPE field, which was used to identify the higher-level protocol associated with the packet. The 802.3 protocol has no TYPE field, but in its place has a two-octet LENGTH field. Thus, it is possible to encounter two Ethernet local area networks that have slightly incompatible formats.

As we learned in the earlier section on IEEE 802 protocols, the 802.3 CSMA/CD protocol, or Ethernet, can support 10Base5, 10Base2, 1Base5, 10BaseT, and 10Broad36 configurations. The choice of configuration further dictates the choice of cabling and equipment necessary to support an Ethernet local area network.

A workstation may be attached to the medium by means of a passive tap, and taps must be separated by multiples of 2.5 meters to reduce signal reflections in the medium. Between the workstation and the tap is the transceiver. The transceiver converts the signals on the medium to a form recognized by the workstation (Figure 6.16).

Wireless Ethernet

One of the major headaches associated with local area networks is installation of the appropriate cabling in the walls, ceilings, and floors. The chore of installing cabling in existing buildings is often difficult and quite expensive. Even if a building is prewired, moving a workstation to another location is often difficult later unless the cabling also moves with the workstation. The **wireless local area network** is a clever solution to the cabling problem. The obvious advantage of a wireless system is precisely what its name implies: there are no wires between workstations. Early wireless networks used infrared signals to transmit data between workstations and servers, but infrared signals cannot pass through walls, floors, or ceilings, so these networks were limited to a single room. More recent wireless networks use omnidirectional high-frequency radio waves that allow stations to be separated by as much as 800 feet in any configuration, with optional capabilities of as much as five miles between workstations. These radio waves can easily pass through walls, and the network supports the Ethernet protocol with data transmission speeds of 2 Mbps.

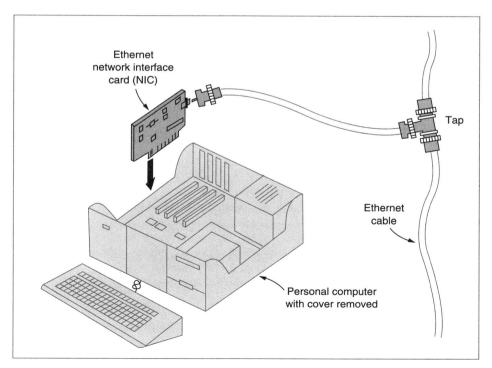

FIGURE 6.16   Ethernet workstation attached to the medium

Fast Ethernet

Fast Ethernet is a variation of Ethernet that can transmit data at speeds up to 100 Mbps over standard Ethernet cabling. At the time this was written, a number of companies were presenting various forms of Fast Ethernet, each one slightly different from the others. Two basic schemes seem to be emerging: keep the 802.3/Ethernet medium access control layer intact, or remove the medium access control layer from the standard.

StarLAN

StarLAN was originally developed by AT&T to provide CSMA/CD access using a star topology. Cabling was provided by unshielded twisted pair cable that could transmit data at 1 Mbps (1Base5). Since its original conception, StarLAN has continued to evolve into an increasingly powerful network. More recent configurations use coaxial and fiber optic cable with transmission rates of 10 Mbps and can support over 1200 users on a single network.

IBM Token Ring

The token ring local area network, which follows the IEEE 802.5 standard, was popularized by IBM. Each workstation attaches to the ring by way of a multistation access unit (MAU), creating the star-wired ring configuration introduced earlier in the chapter.

At the present, two types of token ring network are supported:

1. Type 1 Token Ring is a 16 Mbps network using shielded twisted pair cabling. One unbridged main ring is capable of supporting 32 eight-station MAUs.
2. Type 3 Token Ring is a 4 Mbps network using unshielded twisted pair cabling.

Fiber optic cable can be used on the main ring to interconnect MAUs for greater main ring distances or to connect networks between buildings.

MAP and TOP

MAP and TOP are two examples of industrial-strength local area networks. MAP (Manufacturing Automation Protocol) is a token bus network created by General Motors for use on an automotive assembly line. Actually, MAP defines an entire suite of protocols from layer one up to layer seven of the OSI model and interconnects computers, terminals, and robots. TOP (Technical and Office Protocols) was created by Boeing as a standard for office automation. It too is more than just a local area network as it defines a suite of standards for all seven layers of the OSI model. TOP, however, is built around either an Ethernet network or a token ring network and has not been given as wide a reception as MAP.

AppleTalk

AppleTalk is a local area network created by Apple Computer to support its very popular Macintosh microcomputer. While many other local area network brands support the interconnection of Macintosh computers, AppleTalk was designed primarily for the Macintosh. Its principal characteristics are these:

- The network is CSMA/CD based.
- The network may be configured as a star or bus.
- The Macintosh computer comes cable ready for interconnection to the AppleTalk network.
- The data transmission speed is a relatively slow 230 Kbps.
- Twisted pair, coaxial, or fiber optic cable may be used as the interconnecting medium.
- At most, 32 stations may be interconnected without necessitating additional support equipment.

Because of its limitations, AppleTalk is typically found only on smaller installations of Macintosh computers.

ARCnet

ARCnet, created by Datapoint Corporation, is a popular local area network that does not follow any of the IEEE 802 formats. ARCnet's format is based on a token passing bus and can be configured into a bus topology or a star topology. The original ARCnet transmits data at 2.5 Mbps using coaxial, twisted pair, or fiber optic cable. It is easy to implement, relatively inexpensive, and is suitable for small networks of microcomputers. A newer version, ARCnetPlus, transmits data at 20 Mbps using coaxial cable and is more suitable for larger installations.

ARCnet is often found in the more flexible star topology. Up to 255 workstations are connected to the network through either active or passive hubs (Figure 6.17). An active hub may allow up to 20 interconnections (depending on the model) to workstations or

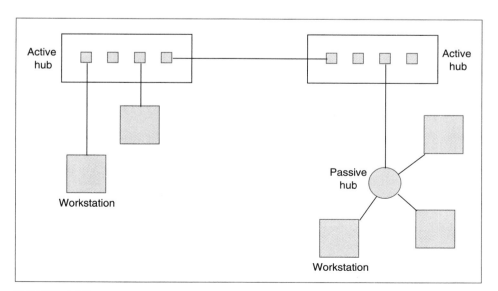

FIGURE 6.17   Star configuration of an ARCnet LAN

other hubs. Active hubs may be chained together and perform a regeneration of the signal passing through. Workstations may be as far as 600 meters from the active hub, and active hubs may be chained together for an overall network span of 6000 meters.

Passive hubs cannot be chained together but can only connect to an active hub and other workstations. Since a passive hub does not regenerate the signal, a workstation may be at most only 30 meters from the hub.

PRIVATE BRANCH EXCHANGE (PBX)

The private branch exchange (PBX) is a privately owned telephone switching system that many businesses and colleges use to provide all their in-house telephone services. Typical functions of a PBX include in-house dialing, dialing outside lines, and most if not all of the common telephone services such as three-party calls, conference calls, auto redial, speed dialing, and phone forwarding. Many of the current systems are also providing voice mail message systems. Because of the increase in digital technology, PBXs are performing all their functions digitally and are beginning to offer services rivaling a local area network. After all, it is not unusual to find every office in a building wired for phone service. If any phone can literally call any other phone, and if each phone line could share data from a computer workstation along with voice traffic, any workstation would have the capability of communicating with any other workstation. Indeed, PBXs and the associated hardware do exist to allow computer workstations and voice telephones to share a single line (**data over voice**, or **DOV**).

Going one step further, if we add a rack of modems to the PBX and connect each modem to an outside phone line, anyone sitting at a workstation can direct the PBX to connect his or her workstation to a modem that would allow dial-up modem access to the outside world. There is almost no limit to the interfaces possible for a PBX: local area networks, wide area networks, other PBXs, IBM 327x devices, SDLC/SNA links, minicomputers and mainframes, and high-speed phone lines such as T-1 (Figure 6.18).

All this may sound too good to be true, and there are indeed four possible limitations to the PBX, a realization that brings us down to earth:

- In the past the PBX has not been known for very high data transfer rates. Speeds of 9600 bps and 19,200 bps were common for a number of years. However, newer models are capable of achieving transfer rates as high as 10 Mbps, which is comparable to a CSMA/CD local area network.
- If computer data from a workstation is to share the same line with voice, they both have to be either digital or analog. Since computers are already digital and more expensive, the telephone should become a digital device. It would be counterproductive to leave the telephones analog and convert the computer data to analog via modems since modern PBXs rely on digital switching circuits. Digital phones do exist, but the analog counterpart is clearly the more prevalent and less expensive.
- The capability of a PBX to handle a large number of possibly lengthy connections simultaneously is limited. PBX switching circuits are designed for a maximum number of short, voice conversations. This maximum number is always below the total number of phone lines, since it is assumed that not everyone will be talking to someone else at the same time.

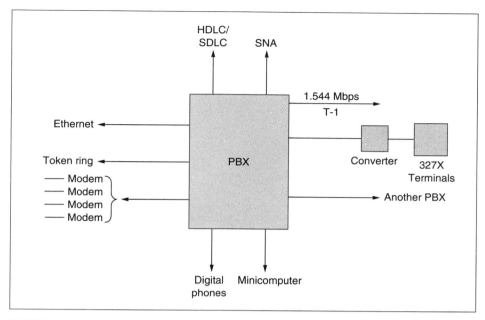

FIGURE 6.18   External connections possible for a PBX

■ A PBX does not possess the ability to broadcast or multicast messages from one station to *n* others.

Despite these shortcomings, a digital PBX can sometimes be a reasonable alternative to installing a possibly expensive resource-consuming local area network.

INTERNETWORKING

Many times a single local area network is not sufficient to support the needs of its users. In these situations, multiple LANs and/or access to a wide area network might provide better service. If this situation is desired, it is possible to interconnect the multiple networks (Figure 6.19). This interconnection is called **internetworking** and a description of it could easily fill several volumes because of the wide diversity of networks.

If you wonder why someone would want to connect two or more networks together, there are several valid reasons:

■ Your company may have a local area network that has reached its designed capacity, but more workstations need to be added to the network. A second identical network could be installed and connected to the first network with a bridging function, allowing all workstations on both networks to intercommunicate.

■ Your company could have two identical local area networks, one for the research department and one for the design department. If you could interconnect the two, it would be possible to share data and resources among all workstations in the two departments.

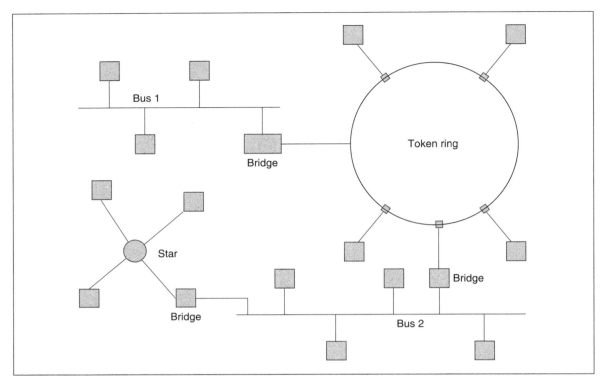

FIGURE 6.19   Interconnection of multiple local area networks

■ Suppose your company has two local area networks but they use different protocols. One is a token ring and the other is a CSMA/CD network. By interconnecting the two dissimilar networks with an appropriate bridging device, it should be possible for the two networks to share data and resources.

■ Your company may want to interconnect a local area network to a wide area network. With this interconnection, users of the local area network can gain access to the wide area network and possibly access any point in the world.

■ Your company might have a LAN that is sufficiently large but is very busy with high demands on a file server and printer. The company would like to divide the network into two LANs, thus dividing up the workload. Dividing the large network could also provide a level of security. If one network requests data from the other network, a monitoring device could limit the flow of information between the two networks. Another reason to split a large network could be environmental. If the company is physically split between two buildings, a network could be located in each building with an interconnection between the buildings.

Fortunately, many devices are available (with more appearing every day) for interconnecting two or more networks. We can classify these devices into three basic categories: bridges, routers, and gateways.

Bridge

A **bridge** is a device that interconnects two similar or not too dissimilar types of networks (Figure 6.20). Stated another way, a bridge connects two local area networks that have the same or similar physical and medium access control layers. Thus, a bridge can interconnect two identical networks, such as two token rings, with little processing power necessary. A bridge can also interconnect a token ring and CSMA/CD networks, performing the necessary conversions between the slightly different MAC layers.

As a packet of data moves across network A and enters the bridge, the medium access control sublayer source and destination addresses are examined. The bridge, using some form of internal logic, determines whether the destination address is intended for a workstation on network A. If the destination is an address on network A, the bridge does nothing more with the packet, since it is already on the appropriate network. If the destination is not an address on network A, the bridge passes the packet on to network B, with the assumption that the packet is intended for a station on that network. Additionally, the bridge can also perform a checksum computation on the packet and set any necessary error bits before retransmitting the packet.

How does the bridge know what addresses are on which networks? Does the bridge even have to know this? To answer these questions, we must further categorize a bridge into three basic types: the transparent bridge, the source-routing bridge, and the source-routing-transparent bridge.

The Transparent Bridge

The **transparent bridge**, designed for CSMA/CD LANs, performs the routing decisions, freeing the source workstation from this task. On installation, the bridge begins observing the addresses of the packets in transmission on the current network and creates an internal

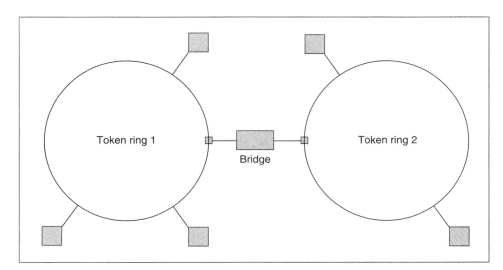

FIGURE 6.20   Two token ring networks interconnected with a bridge

hash table to be used for making future routing decisions. The bridge accomplishes this by using a form of backward learning, or observing where the packet has come from. If a packet is on the current network, the bridge assumes that the packet originated from somewhere on that network. The bridge takes the source address from the packet and places it in an internal table. After watching traffic for awhile, the bridge has a table of workstation addresses for that network. Then, if a packet arrives at the bridge with a destination address other than one in the table, the bridge assumes the packet is intended for a workstation on some other network and passes the packet to the next network.

The Source-Routing Bridge

The **source-routing bridge** was designed for IBM Token Ring LANs. With this type of bridge, the workstation sending the packet has to know the route to the intended destination and inserts into the data packet the appropriate address sequence of network-bridge-network-bridge- . . . and so on. At most, seven bridges may be traversed with one sequence. The bridge is not required to maintain any address tables. It simply has to examine the address sequence and either forward the packet or not. What if the sending workstation does not know the exact sequence of network-bridge addresses? To obtain the proper sequence, the source workstation can dispatch a **discovery frame**. A discovery frame makes its way to the destination, and as it returns to the source, the return path address sequence is recorded and placed in the address field of the packet.

The Source-Routing-Transparent Bridge

The **source-routing-transparent bridge** is a compromise of the first two bridges. Apparently, IBM realized that the CSMA/CD LANs were very popular and were not going to go away. In an effort to combine the transparent bridge of CSMA/CD networks with the source-routing bridge of token ring networks, the source-routing-transparent bridge was created. This bridge normally operates in transparent mode with the backup option of source-routing.

Bridges with Intervening Packet Data Networks

A bridge is also capable of passing a data packet from one LAN to another even if there is a packet data network (such as X.25) between the two networks. In Figure 6.21 a packet leaves workstation 1, travels across local area network X, and arrives at bridge A. Bridge A realizes that the packet is intended for a workstation on network Y and prepares to forward the packet. But when forwarded, the packet must traverse an X.25 network, so the bridge adds X.25 header information *on top of* the existing LAN packet. The X.25 header information is used to get the packet across the X.25 network and to bridge B. At that point, the X.25 header is removed, leaving the original LAN packet, which then travels across network Y to workstation 2.

Sample Operation of a Bridge

As previously mentioned, a bridge can interconnect two local area networks that have similar medium access control layers. Consider the following example of interconnecting a CSMA/CD local area network with a token ring network:

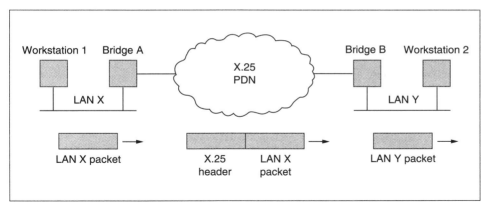

FIGURE 6.21   Two LANs with intervening X.25 packet network

A data packet originates in an upper layer application:

data packet

As the data packet is passed to the logical link control sublayer from the upper layers, an LLC header is added to the front of the data packet:

LLC header + data packet

This LLC data packet is then passed to the medium access control sublayer where the appropriate CSMA/CD MAC sublayer fields are added to the front and rear of the LLC data packet:

MAC header + LLC header + data packet + MAC trailer

This MAC data packet is then passed to the physical layer of the CSMA/CD LAN, which transmits the packet over the medium to the bridge. The bridge, realizing that the CSMA/CD data packet is intended for a workstation on the token ring LAN, strips off the MAC fields leaving the original LLC data packet:

LLC header + data packet

The bridge converts the appropriate CSMA/CD MAC field values to token ring MAC field values, and adds to the front and rear of the LLC packet the appropriate token ring MAC fields:

MAC header + LLC header + data packet + MAC trailer

This new MAC data packet is then sent to the physical layer of the token ring LAN and is transmitted to the receiving workstation. The receiving workstation then strips off all MAC fields:

LLC header + data packet

and LLC fields, leaving the original data packet:

data packet

Router

The second network-connecting device is the **router**. It is a more elaborate device than the bridge. Its primary function is to route data packets between two networks employing protocols that may or may not have medium access control (MAC) layers. An example might be interconnecting a CSMA/CD LAN to the fictitious WideNet network. The WideNet network uses a protocol that has physical, data link, and network layers, but no MAC layer. To perform the routing, the router cannot just look at the MAC layer addresses, because WideNet doesn't have a MAC layer. Instead, the router must dig further into the data packets for more information. Since both the CSMA/CD network and the WideNet network have a network layer (with network layer addresses), the router examines the network layer addresses and uses them to perform the routing. The by-product of digging deeper into the data packets for network routing information is increased processing time, making the router a slower device than a bridge.

Example of a Router's Operation

The following example is the sequence of events that occurs when a CSMA/CD data packet is converted by a router to a WideNet data packet:

A data packet originates in an upper-layer application:

data packet

The data packet is passed to the network layer of the CSMA/CD network and the appropriate network layer header and addresses are added to the data packet:

network layer header and addresses + data packet

This packet is now passed to the LLC sublayer of the CSMA/CD network and the appropriate LLC header is added to the packet:

LLC header + network layer header and addresses + data packet

Next the packet goes to the MAC sublayer of the CSMA/CD network and the appropriate MAC fields are added to the packet:

CSMA/CD MAC header + LLC header + network layer header and addresses + data packet

The packet is given to the physical layer, which transmits the packet over the CSMA/CD local area network to the router, which removes all the CSMA/CD MAC fields:

LLC header + network layer header and addresses + data packet

The router then removes the LLC header:

network layer header and addresses + data packet

The router examines the network layer header and addresses, and creates the appropriate address field(s) for transfer over the WideNet network. These address fields are then added to the data packet:

WideNet header + network layer header and addresses + data packet

The data packet is given to the data link layer of the WideNet network, which applies the appropriate data link fields:

WideNet data link fields + WideNet header + network layer header and addresses + data packet

The packet is transmitted across the WideNet to the intended receiver's physical layer, and then passed to the receiver's data link layer, which removes the WideNet data link fields:

WideNet header + network layer header and addresses + data packet

The data link layer passes the packet to the network layer, which removes the WideNet header, and network layer header and addresses:

data packet

The original data packet is finally passed to the upper layers of the application at the receiving station.

Gateway

The last network-connecting device we examine is the **gateway**, a sophisticated hardware/software device that can convert the protocol of one network to the entirely different protocol of a second network. An example of this would be interconnecting a CSMA/CD LAN with IBM's SNA network. As we saw earlier, the CSMA/CD protocol follows the seven layer OSI model; but an SNA network does not follow the OSI model, instead having its own layers. To accomplish the interconnection, the gateway has to go all the way up to the top layers of the two models, since there are few or no corresponding layers between the two network models. Another name often used for a gateway is a **protocol converter**.

LAN SOFTWARE CONCEPTS

A local area network is of little use without a network operating system. Like a mainframe operating system, a network operating system provides user access to the system and properly manages the network resources. Typically, every workstation on the network requires some level of network software so that the workstation may access the local area network. The network software operates in addition to or in place of the microcomputer operating system.

To help you understand further the operation of a local area network operating system, we have listed below the principal functions a typical system will perform:

- Provide workstation access to a network mass storage device (file server)
- Allow workstations to send print jobs to one or more network printers (print server)
- Provide periodic tape backup of important files on the file server
- Provide an appropriate degree of fault tolerance so that a hardware failure will not cripple the system (such as providing a duplicate hard disk server that "mirrors" the primary hard disk server)

- Support a variety of workstations, types of media, and optional configurations
- Provide for expandability and system/component evolution
- Provide diagnostic and maintenance tools to debug system errors, locate faulty cabling, monitor workstations, generate usage statistics, create user accounts, and provide system security
- Provide a level of transparency so that a workstation user is not required to learn a new set of procedures for accessing programs and data
- Allow concurrent operation of software that is designed for concurrency, and enforce single-user access for software that is not designed for concurrency
- Provide access to other local area networks and wide area networks
- Provide disk caching and file buffering to reduce the number of hard disk accesses

Today there are many local area network operating systems that provide most if not all of the functions listed above. Several of the better-known operating systems include Novell's, 3Com's 3+Open, IBM's OS/2 operating system, and Banyan Vines.

CASE STUDY

To continue with our case study, we now learn that staff in the computing center at XYZ University have decided to replace the computer faculty mainframe terminals in Building B and the computing center support personnel mainframe terminals in Building A with microcomputer workstations. So that the workstations may communicate with the mainframe, all workstations will be interconnected through a local area network. The staff also decided that the library would benefit from connection to a network, thus the local area network will extend through the tunnel between Building B and the library (Figure 6.22).

The staff of the computing center must answer many questions before they can decide on the type of local area network to install. From the information introduced in this chapter, we have formulated a set of questions/criteria that the staff should consider to make an informed decision.

1. How many workstations will initially be interfaced to the network?

At the present moment, XYZ University's computing center staff is considering replacing the six terminals for the support personnel and the seven terminals for the computer faculty. A local area network with 13 workstations does not seem unreasonable. Practically any brand of local area network should be able to support this number of workstations and still operate with reasonable efficiency.

2. What expansion is envisioned in the near future?

This question is always difficult to answer. Many people planning the installation of a new system often underestimate its growth in the near and distant future. The computing center staff, trying to avoid this pitfall, envision a day soon when all faculty members will want a workstation with access to the campus local area network plus access to one or more outside wide area networks. They also anticipate the need soon for a number of microcomputer laboratories for student and classroom use.

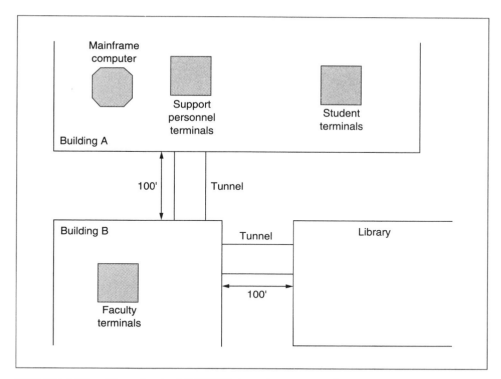

FIGURE 6.22   Hypothetical XYZ University campus layout

Although many schools keep the student and classroom microcomputer labs separate from the university and faculty networks, claiming security reasons, the real reason for the separation is often that the university network cannot support the demand of the additional workstations.

The library is another prime candidate for interconnection to the university network. Students, faculty, and staff working at the library (or in their own offices) would like to have access to one or more electronic card catalogs. If these card catalogs are off campus, the off-campus connection is typically housed within the computing center; therefore, the library needs a network connection to the computing center.

It is apparent that the computing center staff foresees a campus where all students, faculty, and staff will have access to the mainframe computer and a campus-wide network. One possible solution for a campuswide interconnection is to create multiple local area networks within the university, all tied together by means of one high-speed network. For example, the library could have one local area network, the faculty in Building B would use a second local area network, and the support personnel in Building A could use a third local area network. All building local area networks can then be interfaced to one high-speed local area network, or **backbone network**, which would run the length of the campus. If the backbone network is capable of supporting a large amount of traffic, each building could support one or more local area networks that interconnect to the backbone network with bridges.

One advantage of separate local area networks connected to a backbone is security. Users on one local area network would not automatically have access to the local area network of another building, so computing center staff members' files or faculty members' files would not be open for access from other networks on campus. A second advantage is the potential for network growth. If all workstations do not directly address the same network, there is less chance that the one campus network will suffer serious congestion problems.

A clear disadvantage to multiple networks tied together on one backbone network is the cost of installation and maintenance of each separate network. A number of staff members will need to be knowledgeable in supporting all the various networks, including their software and hardware. Backup parts will be necessary for each type of network. Separate maintenance contracts may also be necessary, one for each type of network.

An alternative to providing campuswide network access is to install one local area network that is capable of supporting all users on campus at present and in the near future. While this solution might simplify design choices, reduce maintenance costs, eliminate the need for bridges, and simplify support needs, the questions of security and future expansion are possible negative issues.

The computing center staff has decided to implement the multiple local area networks/high-speed backbone alternative. Therefore, they will first consider one high-speed local area network for the campus backbone, and then multiple smaller local area networks for the individual buildings.

3. Do the users of the workstations require a response time determined by real-time needs?

If the users of the microcomputer workstations required a response in a deterministic amount of time, such as the response that would be necessary to control real-time applications or assembly line production, the best choice might be a medium access control protocol that produces such responses. Since this is not the case at XYZ University, the staff could choose either a deterministic medium access control protocol or a nondeterministic medium access control protocol.

4. What type of traffic will the network carry? Data? Video? Audio?

Originally, the staff at the computing center could see the need for transmitting only data traffic over the various local area networks on campus. If this were to remain unchanged, all local area networks are capable of transferring data between any two points at data transfer speeds between 1 and 100 Mbps. Now, however, many network applications incorporate video and audio into the presentation of simple data. Online encyclopedias are not only capable of displaying the text for Martin Luther King, Jr.'s "I Have a Dream" speech but can also display a video picture of King as well as digitally produce the speech over a speaker. If audio and video presentations such as these are to be transmitted over a local area network, the network must possess a bandwidth wide enough to support the increase in volume of data when this is demanded.

5. How much traffic will the network be required to carry at any one moment?

If the computing center chooses the design in which each building has its own network connected to a high-speed backbone campus network, the number of simultaneous network users on campus is not as important as it would be if there were only a single network on campus. Also, since most faculty, staff, and student network workstations will also be stand-alone microcomputers, the computing center staff anticipate the majority of computing to be performed at the individual workstation, with an occasional file transfer from a network server or from the host mainframe. This assumption is, at best, a guess, and may come back to haunt the staff in the future.

6. Will the choice of workstation dictate the choice of medium access control protocol/network, or vice versa?

If the individual workstations were specialized units that demanded unique interfaces or ultra-high-speed data transfers, it might be necessary to consider a specialized network or unique medium access control protocol, such as an FDDI network. Since the majority of campus workstations will be involved primarily with electronic office functions such as E-mail and file transfers, a particular medium access control protocol or network is not required.

7. Does the campus layout dictate a particular topology?

The campus buildings appear to be laid out in a linear fashion, thus suggesting a bus topology as a logical choice. The bus could begin in Building A near the mainframe computer, pass through the tunnel connecting Building B, and continue through the second tunnel connecting Building B with the library. Each building could then connect its workstations with the bus, or each building could contain its own local area network which would in turn interface with the bus.

A ring topology could also be used, however, as Figure 6.23 demonstrates. A star topology is another option, but we would assume that the computing center would be the likely location for the central hub, and it is not located near the center of the campus.

Conclusion

Having addressed these seven questions, the computing center staff have decided to incorporate a bus backbone network running the length of the campus from Building A through the library. Each building will then support one or more local area networks, which will all be interconnected to this backbone network. The individual local area networks will interface to the backbone network by bridges.

The medium access control protocol chosen for the backbone network will be CSMA/CD. This choice was based on several factors:

- CSMA/CD is the most common local area network with plentiful hardware and software at reasonable prices.
- It is not necessary to incorporate a deterministic access method such as token bus because of the intended use of the network.
- CSMA/CD requires relatively simple software.

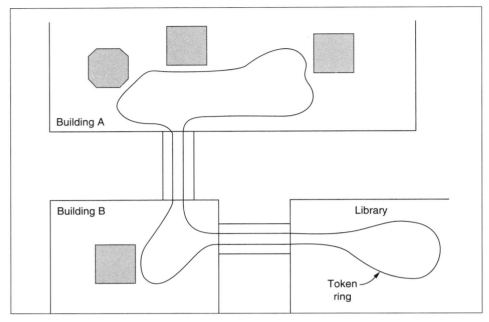

FIGURE 6.23   Campus buildings wired with a ring topology

Given the choice between baseband and broadband signaling, the staff chose broadband signaling because of its high bandwidth and its capability to carry video and audio as well as data.

The bus backbone, however, was not a clear-cut winner. Many members of the computing center staff were in favor of a ring network, such as the 16 Mbps token ring, or even the 100 Mbps FDDI ring. With the new nonfiber FDDI standards emerging and the promise of Fast Ethernet the decision becomes even more difficult.

SUMMARY

A local area network (LAN) is a communication network that interconnects a variety of data communicating devices within a small area and transfers data at high transfer rates with very low error rates. A major advantage of using a local area network is the sharing of data, software, and peripherals. Unfortunately, local area networks are costly and time-consuming to maintain. Several common functions of local area networks include file serving, print serving, electronic mail, process control and monitoring, and teleconferencing.

A local area network can be configured as a bus/tree topology, a star topology, or a ring topology. The bus topology can be further divided into baseband or broadband. A baseband local area network uses digital signaling and has one channel, while broadband local area networks use analog signaling and can support hundreds of simultaneous channels.

For a workstation to place data onto a local area network, the network must have a medium access control protocol. A bus local area network can use either the popular

CSMA/CD protocol or the deterministic token bus protocol. A ring topology can use the popular token ring protocol as its access method. In order to standardize the medium access control protocols, IEEE created the 802 series of network standards.

While there are hundreds of brands of local area networks, the more popular ones include Ethernet, StarLAN, IBM's Token Ring, AppleTalk, and ARCnet. As a possible alternative to a local area network, the private branch exchange, or PBX, offers many of the same services. While it has many advantages of local area networks, some users believe that the PBX falls short in the area of providing multiple, high-speed connections.

Often it is necessary to interconnect multiple local area networks, or to link a local area network with other types of networks. Bridges provide an interconnection at the data link layer of two networks, routers provide an interconnection at the network layer, and gateways provide an interconnection at the higher layers of two networks.

Finally, a local area network is useless without a properly designed network operating system. A local area network operating system provides services such as file and print serving, file backup, and diagnostic and maintenance routines.

EXERCISES

1. What properties set a local area network apart from other networks?
2. List three advantages of using local area networks.
3. List three disadvantages of using local area networks.
4. Given the three common application areas of local area networks, what functions would typically be performed for each application area?
5. State three advantages of baseband signaling over broadband signaling.
6. State three advantages of broadband signaling over baseband signaling.
7. In the IEEE 802.5 token ring protocol, the ED field has two copies of the A bit and the C bit. Why are there two copies? Doesn't the packet checksum include these bits in its checksum calculation? If not, why not?
8. In a star topology, what would be the advantage of keeping the file server separate from the central hub?
9. Explain the statement that a ring is "composed of segments of one-way point-to-point cables strung between pairs of repeaters."
10. If, on a token ring LAN, a workstation is turned off, does the entire network stop? Explain.
11. What do we mean when we say the CSMA/CD medium access control protocol is "nondeterministic"?
12. Using p-persistent CSMA/CD with $p = 0.1$, if 10 workstations are listening to a busy medium, what is the probability that a collision will occur once the medium becomes idle? Explain.
13. Draw a hypothetical token bus configuration with four active workstations. Show the "neighbor list" that each workstation would need to create a logical ring.
14. Explain the difference between 10Base5, 10Base2, 1Base5, and 10BaseT.
15. Suppose station A wants to send the message HELLO to station B on an IEEE 802.3 local area network. Station A has the binary address "1" and station B has the binary address "10." Show the resulting MAC sublayer frame (in binary) that is transmitted. (Don't calculate a CRC; just make one up.)

16. List the advantages of FDDI over a standard token ring network.
17. How are a PBX and a local area network similar?
18. Under what circumstances is a bridge used? A router? A gateway?
19. What is the difference between a transparent bridge and a source-routing bridge?
20. There is also an interconnection device called a "brouter." What do you think its functions are?
21. List four common functions of a local area network operating system.
22. *Solve It Yourself Case Study:* There is a retail department store that is approximately square, 35 meters on a side. Each wall has two entrances equally spaced apart. At each entrance is a point-of-sale cash register. Suggest a local area network solution that interconnects all eight cash registers. Address each of the seven questions posed in the case study above.

CHAPTER 7

WIDE AREA NETWORKS

INTRODUCTION

Local area networks are capable of serving a wide number of users over a small area, such as a room, building, or group of buildings. A wide area network, however, can span a state, a country, and even an ocean. One of the earliest and most famous wide area networks was ARPANET; created in the late 1960s by the U.S. government as a research vehicle, it connected select research universities, military bases, and government labs across the United States. Its use has grown tremendously since the network's creation, and the ARPANET has evolved into the present-day Internet.

A second common example of a wide area network is the public data network provided by a number of private companies. A person or company wishing to transmit a volume of data over a wide distance purchases an agreement with a public data network company for an agreed-on period of time. The public data network company will then provide the "network" for the transmission of the data. Standards for interfacing with this network have been created, and terminals that connect to the network must follow these standards.

Our third example of a wide area network is IBM's System Network Architecture (SNA). Actually, SNA is a model for IBM's networks, much like OSI is a model for non-IBM networks. In this chapter we introduce some of the terminology and basic concepts of SNA and compare the SNA model with the OSI model.

A fourth example of a wide area network is the set of standards that make up the Integrated Services Digital Network (ISDN). Although ISDN was designed during the 1980s, it has not yet achieved widespread acceptance. Thus, discussion of ISDN is deferred until Chapter 11, "Emerging Technologies."

PUBLIC DATA NETWORKS AND X.25 _____

Data transfer over short distances can be adequately handled by local area networks (LANs), metropolitan area networks (MANs), and even dial-up modem transmission. But when the distance covered encompasses a state or a country, LANs and MANs can no longer do the job. Dial-up long-distance transmission using voice-grade telephone lines and modems has improved in quality and reliability over the years, but it is still plagued by transmission impairments and high telephone line costs. One possible solution, presented by a number of companies, is to provide a user with a local connection to a long-haul network for a fee. In this way, the "hired" company deals with the details of transmitting the data across the network. As shown in Figure 7.1, the user at location A (DTE) transmits its data to network host (DCE) 1. The network has the responsibility of transmitting the data across the subnet to the destination DCE, where the user at site B (DTE) receives the information. Networks such as these are termed **public data networks**, or **PDNs**.

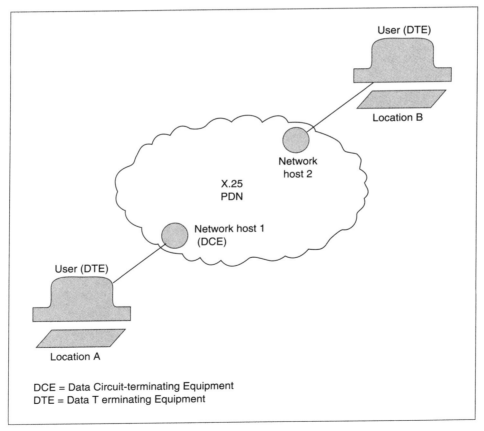

FIGURE 7.1   Two users accessing a public data network (PDN)

As an analogy, a PDN is similar to a package delivery system such as the United Parcel Service (UPS). A person takes the package to the UPS depot, and UPS ships it across the country in the best manner possible. The person sending the package does not necessarily care how the package gets there, just as long as it arrives in a reasonable length of time and is not crushed in the process. Similarly, a user of a PDN does not care how the network gets the "packet of data" to the intended receiver, just as long as it arrives in a reasonable length of time and contains no errors.

Another advantage of using a PDN is that User A pays for only the number of characters of data transferred, not for the total time a connection is established between sender and receiver. This is analogous to talking on the telephone to a friend and being charged only for the number of words you speak, not the overall connect time. As we know, the phone system does not currently operate in this manner.

So that a user may "connect" its DTE terminal to a PDN network, CCITT created the **X.25** standard in 1974, which sets a uniform approach for connecting to a network station. The use of X.25 involves a high-speed synchronous interface and requires a fair amount of software and computing power; therefore, when X.25 is used to connect to a PDN, it requires more than a simple dumb terminal. If a user does not have a workstation with sufficient power, or is using a low-speed asynchronous DTE, X.25 cannot be used. Luckily, CCITT has provided other means to connect low-speed, asynchronous terminals to a PDN. They are discussed later.

The Three Levels of X.25

X.25, similar to the OSI model, is divided into levels. However, since X.25 originated before the OSI model, the three levels of X.25 do not fit precisely into the lower three layers of the OSI model. The lowest level of X.25, the **physical level**, follows the X.21 standard introduced in Chapter 3, but X.21 does not have to be the physical level protocol used. Almost any physical layer protocol could be used, since, like most network models, the X.25 levels are separate from one another and do not contain overlapping protocols. X.21 is not widely used around the world, and X.25 allows the use of EIA-232 or the V.24/V.28 protocols with a few simple modifications. When you use EIA-232 or V.24/V.28 in place of X.21, the physical level interface is termed **X.21 bis**.

The second level of X.25, the **data link level**, is responsible for creating a cohesive, error-free connection between the user's DTE terminal and the network DCE station. These responsibilities are virtually identical to the data link layer of the OSI model. X.25 allows two variations on the HDLC data link protocol, LAP (Link Access Protocol) and LAP-B (Link Access Protocol-Balanced). For now, picture HDLC (or SDLC) since the LAP and LAP-B standards are very similar to those of HDLC. When a data packet is created at the packet level (the third level, immediately above the data link level) and passed to the data link level for transmission across the network, the packet-level data packet becomes the data field for the LAP (HDLC) frame and is encapsulated with beginning and ending 8-bit flags, control field, address field, and frame check sequence (Figure 7.2).

Note that the LAP fields (Flag, Control, Address, Frame Check Sequence, and Flag) remain with the data packet only for the transmission from DTE to DCE. Once the frame arrives at the DCE, and before it is placed onto the network, all LAP fields are removed.

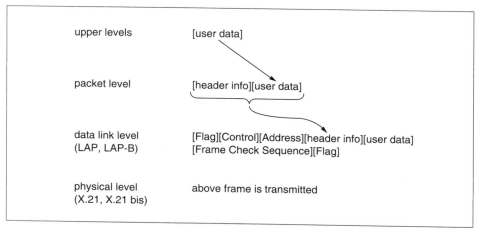

FIGURE 7.2   X.25 levels and the flow of data between levels

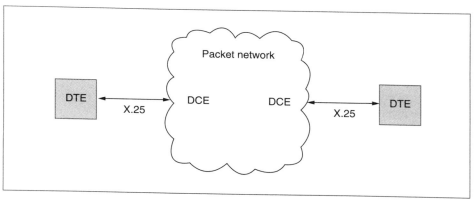

FIGURE 7.3   X.25 connection of user to packet network

The third level of the X.25 protocol is termed the **packet level**. This level deviates the most from the OSI's third layer, the network layer. The packet level's main responsibility is to establish a connection between the user (DTE), the network, and finally the receiver (DTE) at the other end of the circuit (Figure 7.3). Because it is possible for a user to establish multiple connections simultaneously, each network connection is identified by a 4-bit logical group number plus an 8-bit logical channel number. While all 4096 possible connections are not used, there is sufficient room for many concurrent logical connections over each physical channel.

Four Interface Options of X.25

A user of an X.25 PDN has four interface options from which to choose. A user may establish a **permanent virtual circuit**, which is quite similar to a leased line in a public telephone network. Thus, before a session begins, the two users (sender and receiver) and the network administration reach an agreement for a permanent virtual connection, and a logical connection number is assigned. This assigned logical connection number is then used for all future transmissions between the two users.

A second type of interface is the **virtual call** and is similar to the standard telephone call. The sending or originating DTE issues a *call request* packet, which is sent to the network. The network routes the call request packet to the destination DTE, which can accept or reject the request for a connection. If the destination DTE accepts, a *call accepted* packet is sent to the network, which transposes the packet into a *call connected* packet. The call connected packet returns to the originating DTE and a virtual circuit is established. The two DTEs are now in a data transfer state until either DTE issues a *clear request*.

The above two interface designs are connection oriented, in that they require the creation of a virtual circuit with the consent of both DTEs and the network. This is a logical procedure when a considerable number of data packets will be transferred between the two DTEs. But what if a DTE wishes to send only one packet of information to another DTE? It is wasteful to spend so much time making a connection for only one packet of data. The third connection strategy, the **fast select**, allows one DTE to transfer data to another DTE without the call establishment and termination procedures described earlier. The originating DTE sends a fast select packet to the network. This packet may also contain up to 128 bytes of user data. When the network delivers this packet to the destination DTE, the destination DTE has two response choices. The first choice is to return a *clear request* packet, which may also contain 128 bytes of user data. When the originating DTE receives the *clear request* packet, the message is "I accepted your packet, but that's it; thank you very much." The destination DTE could also respond with a *call accepted* packet, which means "I accepted your packet, and let's establish a connection for more data transfers." At that point, the two stations continue sending and receiving data as if a virtual call has been established.

The final interface option is the **fast select with immediate clear**. This technique is similar to the fast select, but the destination DTE has only one possible response and that is to return a *clear request*. On receipt of the *clear request*, the originating DTE returns a *clear confirmation* to the destination DTE. Thus, if a DTE has only one packet of data to send, this fourth and final interface choice is the simplest.

Packet Types and Formats

Several types of packets exist to help X.25 networks establish connections, transfer error-free data packets, and clear connections. These are identified in Figure 7.4.

The call setup and clearing packets are used as described above for creating a virtual call. The originating DTE issues a *call request* packet, the network delivers an *incoming call* packet to the destination DTE, the destination DTE responds with a *call accepted* packet, and the network delivers a *call connected* packet to the originating DTE. A similar dialogue follows for clearing a call.

The *data* and *interrupt* packets are used for transferring data and interrupting transmission. Similar to HDLC commands, the flow control packets are used to acknowledge or suspend the actions of the DTEs. The *reset* and *restart* packets are used by the

Packet Types

From DCE to DTE		From DTE to DCE
	Call Setup and Clearing	
Incoming Call		Call Request
Call Connected		Call Accepted
Clear Indication		Clear Request
DCE Clear Confirmation		DTE Clear Confirmation
	Data and Interrupt	
DCE Data		DTE Data
DCE Interrupt		DTE Interrupt
DCE Interrupt Confirmation		DTE Interrupt Confirmation
	Flow Control and Reset	
DCE Receive Ready		DTE Receive Ready
DCE Receive Not Ready		DTE Receive Not Ready
		DTE Reject
Reset Indication		Reset Request
DCE Reset Confirmation		DTE Reset Confirmation
	Restart	
Restart Indication		Restart Request
DCE Restart Confirmation		DTE Restart Confirmation
	Diagnostic	
Diagnostic		
	Registration	
Registration Confirmation		Registration Request

FIGURE 7.4   X.25 packet types

network and DTEs to recover from errors, such as loss of a packet, congestion, loss of the network's internal virtual circuit, or sequence number error. The reset packet can be used to re-initialize a virtual circuit, while the restart packet is reserved for more serious calamities.

All the above packet types can be packaged into four different packet formats (Figure 7.5). The two forms of the data packet format are used for transferring data across the network. The interrupt packet format is obviously used for transmitting interrupt packet types, and the control packet format is used for the remaining types of packets.

Several fields merit further explanation. The Group # and Channel # are the combined 12-bit fields that identify a particular logical connection. The P(S) and P(R) fields are essentially identical in form and function to the N(S) and N(R) fields of HDLC and SDLC (Chapter 4). Note that the second level of X.25 (the data link level), which follows the LAP-B protocol, also includes N(S) and N(R) fields. However, the N(S) and

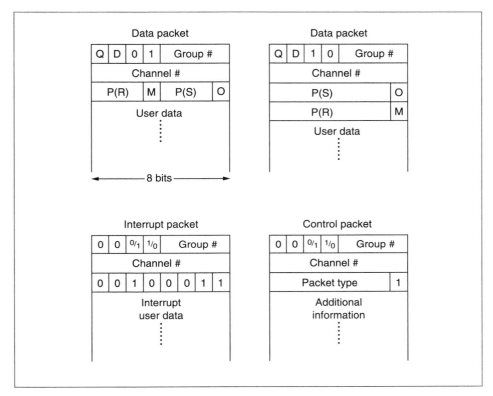

FIGURE 7.5   The four packet formats for X.25 packets

N(R) fields at the data link level are used to control the flow of packets sent between originating DTE and originating DCE. The P(S) and P(R) fields in the third level, the packet level, are used to control the flow of packets between the originating DTE and the network (if the D bit = 0), or between the originating DTE and the destination DTE (if the D bit = 1). Think about this last statement. By providing flow control between the two DTEs, X.25 provides an **end-to-end** flow control. End-to-end flow control is not a function of the OSI network layer, but a function of the OSI transport layer. Thus we see a good example of how the X.25 packet level is not the same as the OSI network layer.

Finally, the Q field is an optional bit and is used to indicate whether the packet contains user data or control information. The M bit tells you when more data packets belonging to the current block of data will follow this packet.

X.3, X.28, and X.29

If you wish to interface a low-speed asynchronous terminal to a PDN, you cannot use X.25. Instead, you must use the combination of X.3, X.28, and X.29 protocols. These three standards combined have been labeled **virtual terminal protocols**, since almost any low-speed terminal should be able to interface a PDN using them.

If a low-speed asynchronous terminal is to interface a PDN, the terminal must connect to a Packet Assembler/Disassembler (PAD). This PAD, which follows the X.3 standard, accepts the low-speed asynchronous data from a terminal and transforms it into the packet format necessary to traverse a PDN. The PAD can be at the same location as the terminal or at the node site of the PDN. If the PAD is located with the terminal, the communication line from terminal/PAD to PDN follows X.25 (Figure 7.6). If the PAD is located with the PDN node, the communication line from terminal to PAD/PDN follows the X.28 standard. Finally, the circuit from PAD to host follows the X.29 standard. It is not necessary at this time to describe the details of X.3, X.28, and X.29 as they are fairly technical and complicated. Interested readers should consult the reading list at the end of the text for further information on the X.25 family of protocols.

THE INTERNET AND TCP/IP

During the late 1960s a branch of the federal government titled the Advance Research Projects Agency (now the Defense Advanced Research Projects Agency) created one of the country's first wide area packet switched networks, the ARPANET. Select research universities, military bases, and government labs were allowed access to the ARPANET for services such as electronic mail (simple mail transfer protocol, or SMTP), file transfers (FTP), and remote log-ins (TELNET).

In 1983 the Department of Defense broke the ARPANET into two basically similar networks: the original ARPANET for experimental research, and MILNET for military use. While MILNET continues to function as the main network for the military, ARPA-

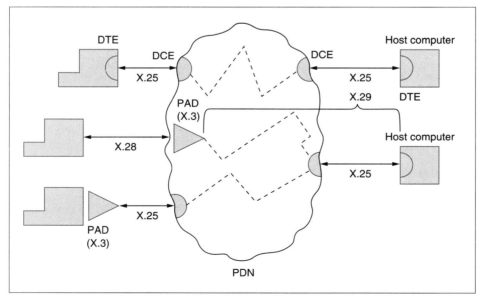

FIGURE 7.6   Terminal interface options for a PDN

NET has been phased out and replaced with new technology. The main impetus to phase out the aging ARPANET occurred in 1987 when the National Science Foundation established a Division of Network and Communications Research and Infrastructure. The goal of this new division was to provide a nationwide network to encourage the exchange of scientific and engineering research. This nationwide network, known as the Internet, is a three-level hierarchy consisting of a high-speed, cross-country **backbone**, a set of regional or midlevel networks that connect to the backbone and each span a small geographic area, and a set of access or "campus" networks that connect to the midlevel networks. The NSFNET backbone, as of 1990, is shown in Figure 7.7. At the present time the Internet is estimated to interconnect $\approx 10^4$ networks and $\approx 10^8$ computers.

To allow the interconnection of so many types of computers and networks, the Department of Defense created a set of protocol standards or services you can use to interface a computer network to the Internet. As shown in Figure 7.8, the services are layered so that a user of the Internet requires the use of one or more application services, which require the use of the reliable transport service, which requires the use of the connectionless packet delivery service. Note how similar in concept this model is to the OSI seven layer model.

The application services consist of applications such as electronic mail (SMTP), file transfer (FTP), and remote log-in (TELNET).

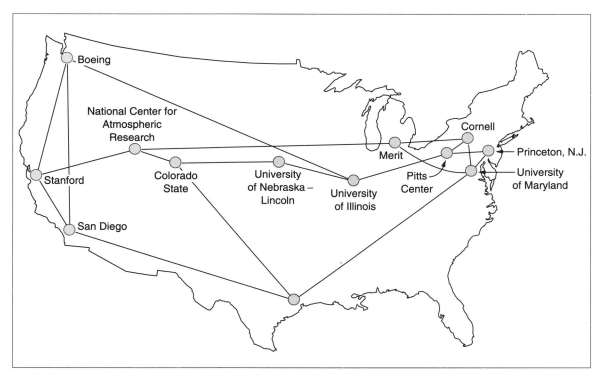

FIGURE 7.7   **NSFNET backbone as of 1990**

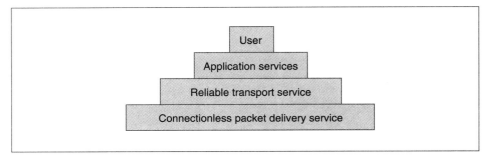

FIGURE 7.8   Set of protocol standards created by the Department of Defense
for interfacing Internet

The reliable transport service turns an unreliable subnetwork into a reliable network, free from lost and duplicate packets. The software that provides this service is called **TCP (Transmission Control Protocol)**.

The connectionless packet delivery service provides an unreliable, connectionless network service in which packets may be lost, duplicated, delayed, and delivered out of order. Even worse, the sender and receiver of these packets may not be informed that problems have occurred in transmission. Thus, the need for a reliable, overlaid transport service is evident, to cover for the possible shortcomings of the delivery service. This connectionless packet delivery service is called **IP (Internet Protocol)**.

Since TCP/IP has proven quite effective for the Internet, many universities and corporations use it for the internal interconnection of heterogeneous networks. TCP/IP is essentially two sets of standards: the TCP set, and the IP set; therefore, we will examine each set independent of the other. TCP, which relates closely to the transport layer of the OSI model, is covered in Chapter 9.

The Internet Protocol (IP)

The Internet Protocol (IP) provides a connectionless data transfer over heterogeneous networks by passing and routing packets of data called IP datagrams. To accomplish this, all packets that are passed down from the transport layer are encapsulated with an IP header that contains the information necessary to transmit this packet from one network to another. If you examine Figure 7.9 in detail you can better understand how this header information is applied. Station A creates a packet of data and passes it to the transport layer (TP) of the software. The transport layer passes the packet to the IP layer, which attaches an IP header (Figure 7.10) to the transport layer packet creating an IP **datagram**. Next, the appropriate LLC and MAC layer headers are encapsulated over the IP datagram and the datagram is sent to Gateway 1 via LAN1. Since Gateway 1 interfaces LAN1 to Brand X wide area network, the MAC and LLC layer information is stripped off, leaving the original IP datagram. At this time the gateway may use any or all of the IP header information to perform the necessary internetworking functions. The necessary Brand X wide area network information is applied and the packet is transmitted over the wide area network to Gateway 2. When the packet arrives at the Gateway 2, the Brand X network information is stripped off, leaving once again the IP

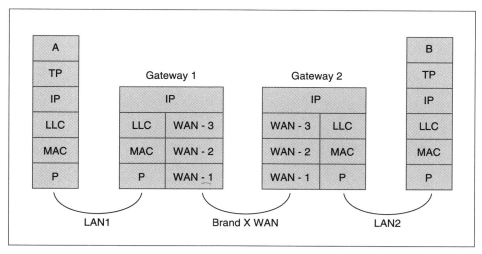

FIGURE 7.9   Progression of a datagram/ packet from one network to another

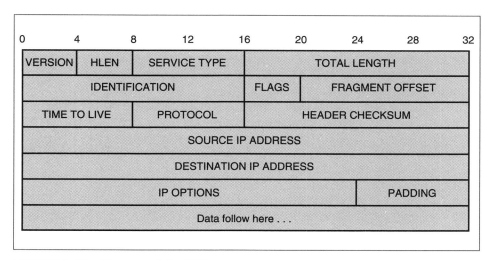

FIGURE 7.10   Format of the IP header

datagram. The appropriate LLC and MAC header information is then applied for transfer of the packet over LAN2, and the packet is transmitted. On arrival at Station B, *all* header information is removed leaving the original data packet.

Each gateway, when it has the packet in IP datagram form, may make several decisions affecting the future of the datagram.

■ The gateway makes routing decisions based on the address portion of the IP datagram.

■ The gateway may have to fragment the datagram into smaller datagrams if the next network to traverse has a smaller maximum packet size.

■ The gateway may determine that the current datagram has been hopping around the network for too long and it may delete the datagram.

Each of these functions is examined in more detail later in the chapter.

IP Header Format

Figure 7.9 demonstrated how an IP header was attached to the upper-layer data packet, creating the IP datagram. This IP header information can then be used to make routing decisions at every gateway. To learn the kind of information that is placed in an IP header, examine Figure 7.10, which shows in detail the individual fields of this header. As you explore each field, you will discover how IP works and why it is capable of interconnecting so many different types of networks.

The Version field contains the version number of the IP protocol being used, just in case a new version becomes available. We don't want two users using two different versions of the IP protocol format at the same time.

The HLEN (*Header LEN*gth) field contains the length of the datagram header measured in 32-bit words. Since the two fields *IP Options* and *Padding* may be of variable length, the receiver must know when the datagram header ends and the datagram data begins.

The Service Type field allows the sender of the datagram the option of specifying the importance level of the datagram (ranging from normal precedence to network precedence). Unfortunately, most gateway software ignores this field at the present time. The *Total Length* field, as its name implies, specifies the total length of the datagram, header plus data.

The next three fields, *Identification, Flags,* and *Fragment Offset* are used to fragment a datagram into smaller parts. Every network has a maximum size packet that it can transfer through its circuits. This maximum size is controlled by network hardware, software, or other factors. Since the IP protocol is designed to work over practically any type of network, it must also be able to transfer datagrams of varying sizes. Rather than being limited to the smallest maximum packet size in existence (who even knows what that is?), the IP protocol allows a gateway to break or fragment a large datagram into smaller fragments that will "fit" onto the available network.

As an example, consider the IP datagram shown in Figure 7.11(a). The data portion of the datagram is 960 bytes (octets) in length. To be transmitted over a particular network, each data portion of a datagram can be no larger than 400 bytes, thus the original datagram needs to be fragmented into three blocks of 400, 400, and 160 bytes, respectively. Therefore, three fragments are created, each with an almost identical IP header. The *Identification* field contains an identification that is unique to these three fragments and is used to reassemble them into the original datagram. The *Fragment Offset* field contains the offset, in units of eight bytes, from the beginning of the original datagram data field. Note, in Figure 7.11(b), the offsets are 0, 50 ($50 \times 8 = 400$), and 100 ($100 \times 8 = 800$). Within the *FLAGS* field is a *more* bit, which indicates whether there are more fragments of this original datagram following.

An additional bit in the Flags field is the DON'T FRAGMENT bit. If set to 1, it indicates that this datagram cannot be fragmented. This field may be used for special control datagrams that would not function correctly if only a part of the original datagram arrived at a destination. If a gateway determines that a datagram must be fragmented,

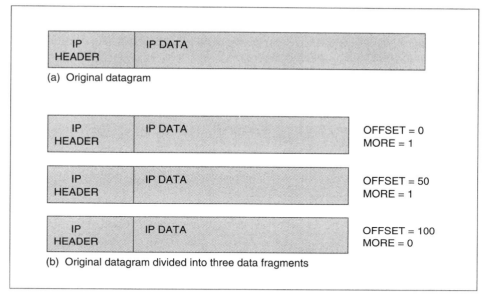

FIGURE 7.11   Fragmentation of an IP datagram into three fragments

but the DON'T FRAGMENT bit is set, the datagram is discarded and an error message is returned.

The *Time to Live* field indicates how long this particular datagram is allowed to live within the system (bounce from network to network). Each gateway along the route from source to destination decrements the *Time to Live* field by 1. Also, each gateway determines how long that datagram has been sitting in gateway queues waiting for service, and decrements the count by that value. When the value of the field reaches zero, the gateway deletes the datagram and sends an appropriate error message.

The *Protocol* field is an integer value that indicates the format used to create the data portion of the datagram. The mapping of integers to formats is controlled by a central networking authority.

The *Header Checksum* field is an arithmetic checksum that is applied to the fields of the IP header only and not to the data portion of the datagram. Since the data portion of the datagram essentially depends on the system transmitting the message and is not a concern of IP, an upper layer is responsible for inserting a checksum (if desired) into the data portion of the datagram.

The following two fields, *Source IP Address* and *Destination IP Address*, contain the 32-bit IP initial source and final destination addresses of the datagram. There are five possible forms for an IP address, shown in Figure 7.12.

Each IP address has two parts: a NetID that indicates a particular network, and a HostID that indicates a particular host on that network. A Class A address allows for 2^7 NetIDs and 2^{24} HostIDs and is usually reserved for very large networks. Class B addresses allow for 2^{14} NetIDs and 2^{16} HostIDs, while Class C addresses allow for 2^{21} NetIDs and 2^8 HostIDs (smaller networks). A particular address class can be used depending on the configurations of networks and hosts. Class D addresses are available

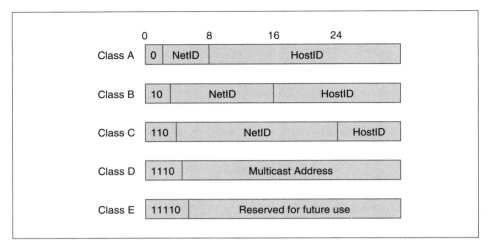

FIGURE 7.12   Five formats of an IP 32-bit address

for those networks that allow multicasting of messages (transmitting to more than one host at a time).

Most users of the Internet are more familiar with the dotted decimal notation of an Internet address, such as 149.164.187.1. This dotted decimal notation is simply the 32-bit address broken into four octets and then written as four decimal values: $(10010101\ 10100100\ 10111011\ 00000001_2 = 149\ 164\ 187\ 1_{10})$.

The *IP Options* field is of variable size and contains the options (if any) pertaining to a particular datagram. Three of the more interesting options used are the **record route option**, the **source route option**, and the **timestamp option**. The record route option, when enabled by the sender of the datagram, causes each gateway from source to destination to record its address in the option field. Thus, when the datagram arrives at the destination, the receiver has a record of each gateway hop along the datagram's path.

The source route option allows the sender of the datagram to specify a particular path that the datagram is to follow through the Internet. Since the average user has no idea or little interest in what route a message takes, this option is generally used only by network administrators and advanced users.

The third option, the timestamp option, is similar to the record route option in which each gateway along the datagram's path records both the gateway address and a 32-bit integer timestamp.

The final field of the IP header, the *Padding* field, simply fills the space after the variable size IP Options field with zeroes to ensure that the IP Options + Padding fields are an exact multiple of 32 bits.

IP Errors and Control Messages

Problems can develop when datagrams are transmitted throughout the Internet. Time to Live violations, the need to fragment a *don't fragment* datagram, and checksum errors are just three examples of conditions that can generate an error message. If an error condition arises and an error message is needed, who creates it and what does it look like? Along with the IP software protocol, there is another software protocol designed to handle error and control messages: **Internet Control Message Protocol (ICMP)**. All gateways have this software. If an ICMP message is required, the gateway creates an ICMP message, which consists of the ICMP Header followed by the ICMP Data, as shown in Figure 7.13(a). Note that the ICMP message, which consists of its own header and data, becomes the IP Data field with an IP Header encapsulated in the front of that message.

The format of the ICMP Header varies, depending on the type of message being sent. However, all types of header formats start with three common fields:

- *Type*: identifies the type of ICMP message
- *Code*: provides further information about the message type
- *Checksum*: performs an arithmetic checksum on the ICMP Header

The different types of ICMP messages are listed in Table 7.1. They are not discussed here as most are self-explanatory.

The ICMP Data portion of the message contains the IP Header and the first 64 bits of the IP Data field whenever the ICMP message is reporting an error condition.

Whenever an ICMP message is generated, it is always sent to the original source of the datagram. If a destination is unreachable, the source is informed that the datagram did not reach its intended destination. But if a routing problem exists halfway through the transmission of a datagram and an error message is returned to the source, it is unlikely that the source can do much about the error, since the source has no idea which route the datagram followed once it departed the source gateway (unless the source route option was employed).

IBM's Systems Network Architecture (SNA)

Most computer companies, in order to stay competitive in the marketplace, offer more than one model of computer or computer peripheral. No one, however, matches the

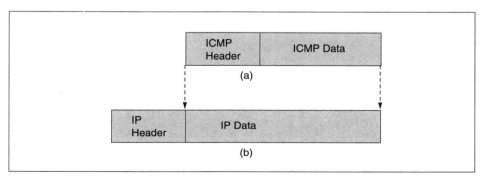

FIGURE 7.13   **ICMP** message format

TABLE 7.1 Different Types of ICMP Messages

Type Field	ICMP Message Type
0	Echo reply
3	Destination unreachable
4	Source quench
5	Redirect (change a route)
8	Echo request
11	Time exceeded for a datagram
12	Parameter problem on a datagram
13	Timestamp request
14	Timestamp reply
17	Address mask request
18	Address mask reply

volume and variety of IBM, with hardware such as mainframe computers, minicomputers, microcomputers, front end processors, cluster controllers, terminals, printers, tape drives, and disk drives. While this diversity of equipment was an advantage to marketing, there were also disadvantages. A major one was the vast number of communication programs used to interconnect model X computer with model Y peripheral. IBM realized that to avoid drowning in disparate communications software, a single model was needed for designing and implementing computer networks. Thus, in 1974, IBM introduced **SNA (Systems Network Architecture)**. Similar to the OSI model, SNA is a model that describes *how* communications/network software should be created. IBM has continued to revise and update SNA since its introduction in 1974.

Due to its flexibility, the SNA model can support many types of communication environments. These include the following:

- Single host mainframe with local terminals connected via cluster controllers (Figure 7.14a)
- Single host mainframe with local terminals connected via cluster controllers and remote terminals connected via communications controllers (Figure 7.14b)
- Multiple remote host mainframes interconnected via communications controllers and connected to remote terminals (Figure 7.14c)

Hardware Node Types

All SNA networks consist of both hardware and software components. A **node** is defined as the hardware and its associated software components that implement SNA functions. SNA defines three basic types of nodes:

1. **Host subarea node:** This node, which consists of a main processor and a telecommunications access method, controls and manages a network.

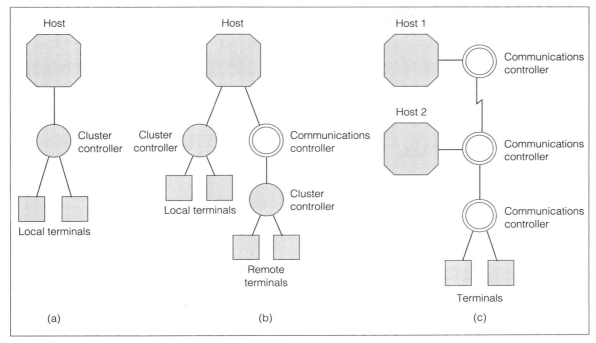

FIGURE 7.14   Three examples of environments supported by Systems Network Architecture (SNA)

2. **Communications controller subarea node:** This node, which consists of a communications controller and a network control program, routes and controls the flow of data through a network.
3. **Peripheral node:** This node can be a cluster controller, a distributed processor, a workstation, or a printer.

Figure 7.15 shows an example of an SNA network with the three different types of nodes.

Network Addressable Units

Each of the above types of nodes contains one or more **Network Addressable Units (NAUs)**. An NAU is a piece of software that allows a user, an application, or a hardware device to use the network. A user, wishing to access the network, connects itself to an NAU. The user/NAU can now address other NAUs to carry out network operations. Main processors, communication controllers, and peripherals must also be connected to NAUs in order to gain access to the network. Thus, an NAU is like a window to the network.

SNA further classifies NAUs into one of three categories: a **logical unit (LU)**, a **physical unit (PU)**, and a **system services control point (SSCP)**. The logical unit is the software that allows a user access to the SNA network. Two end users may communicate with each other by establishing an LU to LU **session**. The end user does

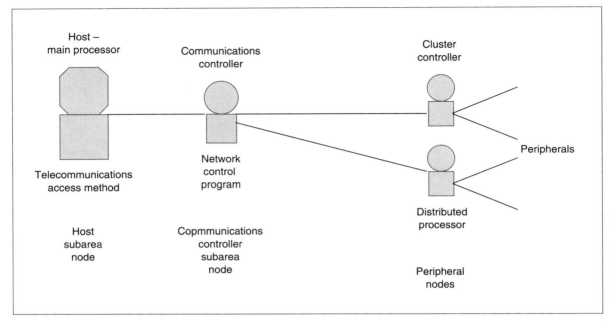

FIGURE 7.15 Sample Systems Network Architecture (SNA) network with three types of nodes

not need to be a human. An end user can be a terminal, a printer, or a software program. SNA defines five basic types of LUs:

LU Type 1: host program to keyboard, printer, card unit, or disk drive
LU Type 2: host program to 3270 display
LU Type 3: host program to 3270 printer
LU Type 4: host program to text processor
LU Type 6: general purpose program to program

Associated with every node is a physical unit, or PU. This physical unit is an administrative-type program that allows the network to bring the node online, perform tests on the node, and perform other similar network management functions. SNA has defined four types of physical units:

PU Type 1: terminals
PU Type 2: cluster controllers
PU Type 4: communications controllers or front end processors
PU Type 5: host computer

The third type of NAU is the **System Services Control Point (SSCP)**, of which there is only one per computer system. The SSCP resides in a PU Type 5 node (host computer). It has control over all cluster controllers, communications controllers, and terminals attached to that host. The **domain** is the collection of hardware and software maintained by the SSCP.

Sessions

To request any services or to transmit any data over an SNA network, a session must first be created. While session creation is an involved process, and following an actual example may be mind boggling, we will attempt to simplify it as much as possible.

SNA defines four types of sessions:

1. **SSCP–SSCP:** This session is created so that one host (SSCP) may communicate with another host (SSCP) with the intent of establishing a session between two users (LUs).
2. **SSCP–PU:** This session is used to allow the SSCP to initialize, control, and stop a PU. In essence, SSCP–PU sessions define the boundaries of the SSCP's domain.
3. **SSCP–LU:** This session is used when an LU first wishes to establish a session with another LU. As we shall see in an upcoming example, an LU first creates a session with its SSCP, so that the SSCP may create an LU–LU session.
4. **LU–LU:** This session is the bread and butter of SNA, used to transmit user data and requests.

As an example, consider the scenario in which a user in one domain wishes to communicate with a user in another domain (Figure 7.16). The precise details of how an SNA session is established vary depending on whether the two users are in the same domain, which user initiates the session, which type of user initiates the session, and whether both users are available at the time of session creation. The following diagram (Figure 7.17) is a simplified example in which the two users are in different domains and both are available. The important point is that for an LU to establish a session with another LU, the initiating LU must ask its SSCP to create the session.

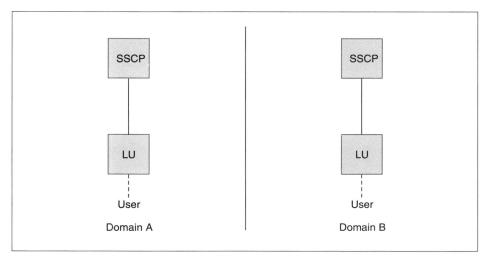

FIGURE 7.16   Two users in separate domains about to create a session

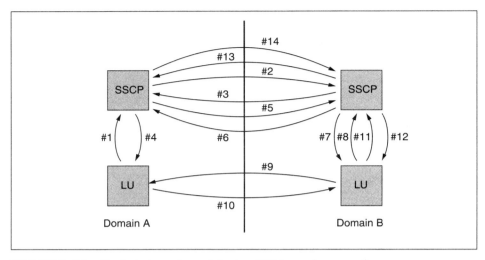

FIGURE 7.17   Packets transferred during SNA session creation

The session creation begins when the LU in domain A (a secondary) sends an *initiate session* packet to its SSCP (packet #1). The SSCP in domain A responds by sending an *initiate cross-domain session* packet (packet #2) to the SSCP in domain B (the primary). SSCP in domain B returns an *acknowledgment* (packet #3) to the SSCP in domain A, which returns an *acknowledgment* to its LU (packet #4). The SSCP in domain A issues an *initiate cross-domain binding* packet to the SSCP in domain B (packet #5) and an *acknowledgment* is returned (packet #6). The SSCP in domain B issues an *initiate binding* to its LU (packet #7) and the LU returns an *acknowledgment* (packet #8). The LU in domain B then sends a *bind* (or session activation) packet to the LU in domain A (packet #9) and the LU in domain A acknowledges (packet #10). The LU in domain B then informs its SSCP of the session binding with the LU in domain A by sending a *session started* packet to its SSCP (packet #11), and the SSCP returns an *acknowledgment* (packet #12). Finally, the SSCP in domain B sends a *session started* packet to the SSCP in domain A (packet #13), and the SSCP in domain A returns an *acknowledgment* (packet #14).

LU 6.2 and PU 2.1

The earlier versions of SNA, like most IBM software, was designed around the primary/secondary concept. A host computer was the primary and all terminals were considered secondaries. A peer-to-peer connection (between two users, two applications, or two programs) did not exist. Thus, LU sessions 0 through 5 were sessions with hardware devices (such as keyboards, terminals, printers, and disk drives). Not until IBM revised SNA to include a new logical unit LU 6.2 was SNA capable of creating a session that allowed two applications (or programs) to talk to each other. LU 6.2, when used in conjunction with PU 2.1, allows the creation of peer-to-peer or program-to-program sessions.

PU2.1 serves an important purpose. Recall from the previous section that to create an SNA session it was necessary to involve the SSCP. Even though it is possible to create an LU 6.2 session without using the SSCP, there are functions the SSCP provides that are necessary for session creation. Thus, PU 2.1 was designed to bring in those necessary session creation functions normally found in an SSCP.

The addition of LU 6.2 was a significant move for IBM and the SNA model as it made possible the interface of non-SNA products or software to an existing SNA network. By modeling their software application after the LU 6.2/PU 2.1 model, a third party could create software that could interact with an SNA network.

SNA Layers

When IBM created SNA as an architectural model for communication systems, it developed an architecture of seven layers. Do not assume, however, that since the OSI model has seven layers and SNA has seven layers the two models are identical. They are fairly similar in one or two layers, but overall they are quite dissimilar. The lowest layer of SNA is the Physical Control Layer and is relatively similar to the OSI Physical Layer. The second SNA layer, the Datalink Control Layer, specifies the use of SDLC as the data link protocol. The third SNA layer, the Path Control Layer, establishes a logical path between source NAU and destination NAU. The Path Control Layer is further divided into three sublayers: the highest sublayer establishes a virtual route from NAU to NAU; the middle sublayer selects an explicit route using the virtual route; and the lowest sublayer divides traffic among parallel communication lines to achieve maximum efficiency.

The fourth layer of SNA, the Transmission Control Layer, creates, manages, and deletes the end-to-end sessions. This layer is similar in function to parts of OSI's transport and session layers. The fifth layer, the Data Flow Control Layer, decides which end of a connection talks next and also performs some error recovery. The sixth layer, the Presentation Services Layer, is a collection of SNA high-level services, such as function management, session control, and presentation features. The final layer, the Transaction Services Layer, is not really a layer but is where an IBM application such as TSO, CICS, or IMS resides.

CASE STUDY

To get some practical experience with wide area networks, we return to the XYZ University computing center. Many students, faculty, and staff at the university have heard much about the Internet and wish to access its electronic mail, remote log-in, and file transfer services. The computing center, wanting to keep its customers happy, has decided to investigate what is required to become an Internet site. State University, which is 150 miles away, is a node for the statewide network and is connected to the Internet. If the XYZ University can establish a connection to State University, it can interconnect to the Internet at that point.

Several steps must be accomplished in order to establish an Internet connection at XYZ University. First, a high-speed line between the two universities is required. There is currently a 14,400 bps line, but the computing center staff have determined that 14,400 bps will be too slow to handle the expected volume of traffic. The university could lease a T-1 line (1.544 Mbps) from the local telephone company, but that might bring much more capacity than the univeristy needs, and could be quite expensive. The computing center has decided to make a few phone calls to learn whether the telephone company can offer a leased line solution between 14,400 and T-1, or if a communications company can offer an appropriate PDN connection. As a third alternative, the computing center may consider one of the newer modem standards that transmits data faster than the currently installed 14,400 bps modem.

Once the communications line is selected, the university will have to buy additional hardware. At a minimum, two routers, one at each end of the new communications line, will be necessary to route Internet data properly between the two universities (Figure 7.18). The university decided to lease a high-quality line from the telephone company and install high-speed modems (57,600 bps) on each end of line.

Now that the physical connection between XYZ University and the Internet has been established, the XYZ network connection will require an Internet network address. So that XYZ University does not choose a network address that is similar to the address of another network somewhere on the Internet, the computing center staff will consult the Network Information Center (NIC), a central authority that assigns the network portion of Internet addresses. Since the XYZ University network is fairly small, a Class C network address is assigned. A Class C network ID allows for 256 host IDs on that one network, a number that should be more than sufficient for the small university.

With a physical network connection established and an Internet address assigned, the XYZ computing center now needs the appropriate software to interface the campus network to the Internet network. This software will follow the TCP/IP model. The IP portion of the model was described in this chapter. Each data packet of an

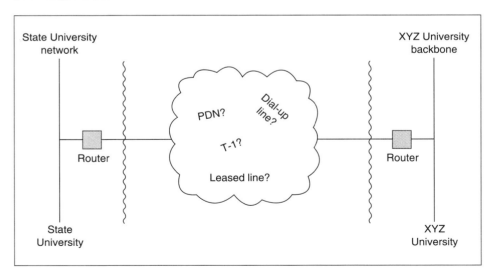

FIGURE 7.18   Proposed Internet connection between State University and XYZ
               University

Internet application, as it is passed down from a higher layer, will be encapsulated with an IP Header that follows the format shown in Figure 7.10. This IP data packet will then be encapsulated with the appropriate network layer headers and data link layer headers for transmission across the link to the State University. Once the packet arrives at the State University site, the data link layer header and network header are removed, and the packet is prepared for transfer across the State University network to the router for the statewide network connection. The packet is then transferred across the statewide network to the site of the Internet connection, and eventually onto the Internet.

SUMMARY

Data transfer over short distances can be adequately served by local area networks (LANs), metropolitan area networks (MANs), and even dial-up modem transmission. But when the distance covered encompasses a state or a country, LANs and MANs can no longer do the job. The wide area network (WAN) is the vehicle of choice for long-distance data transfer. A public data network (PDN) provides a user with a local connection to a long-haul network for a fee. The network has the responsibility of transmitting the data across the subnet from source to destination. To allow a user to "connect" its DTE terminal to a PDN network, CCITT created the X.25 standard in 1974, establishing a uniform approach to making a connection to a network station. If you wish to interface a low-speed asynchronous terminal to a PDN, X.25 cannot be used. Instead, you must use the combination of X.3, X.28, and X.29 protocols.

A second example of a popular wide area network is the Internet. This nationwide network is a three-level hierarchy consisting of a high-speed, cross-country backbone, a set of regional or midlevel networks that connect to the backbone, and a set of access or "campus" networks that connect to the midlevel networks. To allow the interconnection of so many types of computers and networks, the Department of Defense created a set of protocol standards or services to be used in interfacing a computer network to the Internet. The first is a reliable transport service that turns an unreliable subnetwork into a reliable network, free from lost and duplicate packets. The software that provides this service is called TCP (Transmission Control Protocol). The second is a connectionless packet delivery service that provides an unreliable, connectionless network service, or IP (Internet Protocol).

A third example of a wide area network is IBM's Systems Network Architecture (SNA) model. Similar to the OSI model, SNA was designed to provide the architectural blueprints for all of IBM's computer communications networks. It is represented as a seven layer model and can interconnect the wide variety of IBM hardware and software as well as third-party generated software.

EXERCISES

1. How is a public data network similar to a package delivery system?
2. List the three levels of the X.25 standard with a one-sentence description for each level.

3. Compare and contrast the three levels of X.25 with the first three layers of the OSI model.

4. It is possible for X.25 to establish (theoretically) 4096 multiple connections simultaneously. How is this number derived?

5. State the four interface options a user of an X.25 PDN may select, writing one sentence to describe each option.

6. Why would a user desire to create a *fast select* interface on an X.25 network?

7. Explain how the P(S) and P(R) fields of the packet level of X.25 make the packet level of X.25 different from the network layer of the OSI model.

8. List three of the more common application services provided by the Internet.

9. What is the relationship of the TCP and IP protocols?

10. Why does IP have the capability of fragmenting a data packet?

11. Given an IP packet of size 540 bytes, and a maximum packet size of 200 bytes, what are the IP Fragment Offsets and More flags for the appropriate packet fragments?

12. If a packet is bouncing around the Internet for too long and a Time to Live violation occurs, how does the Internet inform the sending station of the violation?

13. List the three basic types of SNA hardware nodes.

14. List the three categories of SNA network addressable units (NAUs).

15. Which type of SNA session is established when an LU first wishes to create a session with another LU?

16. Explain the importance of SNA LU 6.2 to third-party vendors.

17. Compare and contrast the seven layers of the OSI model with the seven layers of the SNA model.

CHAPTER 8

SATELLITE AND RADIO BROADCAST NETWORKS

INTRODUCTION

Satellite and radio broadcast networks are an increasingly vital form of communication in the world of data communications and computer networks. They began with AM radio, FM radio, and television in the 1950s, followed in 1962 by Telstar, the first orbiting satellite. To the present day, dozens of applications have been spawned by this continuously emerging technology. Examples of these applications include long distance television transmission, cellular phones, cellular facsimile (fax), wireless message transmission (pagers), telegraphy, closed caption television (Teletext), burglar alarm systems for home and automobile, emergency radio systems, citizen band radio systems, and global positioning systems for cars that can tell you where you are the next time you get lost.

With this explosive growth of services using satellite and radio systems, the competition for the available frequencies on which to transmit signals is fierce. The Federal Communications Commission (FCC), which controls the allocation of available frequencies within the United States, is constantly considering new alternatives for allowing greater access to the airwaves.

Satellite and radio broadcast networks possess unique properties that set them apart from their wired counterparts:

- Line-of-sight transmission
- Coverage of approximately one-third of the earth's surface by any satellite in geosynchronous orbit (approximately 22,300 miles above the earth) (Figure 8.1)
- Transmissions capable of bouncing off the ionosphere
- Very high transmission rates (as much as 100 Mbps)

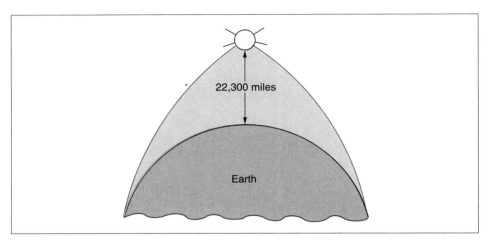

FIGURE 8.1 Satellite in geosynchronous orbit

- Possible long propagation delays due to the long transmission distances
- Fierce competition for the available air waves

Our discussion of these unique systems begins with satellite networks, followed by radio networks and ending with the extremely popular cellular phone networks.

SATELLITE NETWORKS

A satellite communication system is essentially two microwave transmission systems (Figure 8.2). The first system, the uplink, is from a ground station to the satellite; the

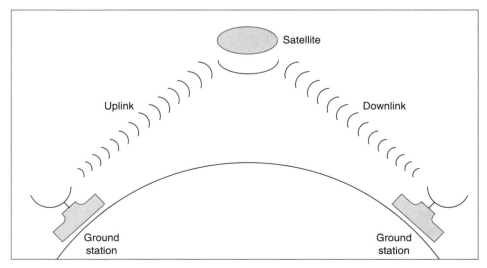

FIGURE 8.2 Simple satellite communication system

second system, after the signal is amplified, is from the satellite down to another ground station (downlink). An example of a satellite communication system is a television relay between Europe and the United States. Many times the average viewer has seen a live interview between two persons on the two continents. The picture quality is often as good as a signal originating from a mile away (unlike the fuzzy satellite broadcasts from the 1960s). The most noticeable feature of the broadcast is the delay when the person in the United States asks the person in Europe a question. This delay is due to the long distance between source and destination and the propagation delay of the signal that results from that distance. There is a minimum of a 0.5 second delay between ground station to satellite and down to ground station. Add in additional hardware and software delays found in connections to the ground stations, and a noticeable delay develops.

Satellite systems can be configured into three basic topologies. These are bulk carrier facilities, multiplexed earth station, and single-user earth station.

Bulk Carrier Facilities

With bulk carrier facilities, the satellite system and all its assigned frequencies are devoted to one user (Figure 8.3a). A satellite system is expensive and capable of transmitting large amounts of data in a very short time; an extremely large application would be necessary to justify economically the solo use of an entire satellite system. An example of such a justifiable system would be a telephone company using a satellite system to transmit thousands of long-distance telephone calls simultaneously. Typical systems operate in the 6/4 GHz bands (6 GHz uplink, 4 GHz downlink) and provide a 500 MHz bandwidth that can be broken further into multiple 40–50 MHz channels.

Multiplexed Earth Station

In a multiplexed earth station, the ground station accepts input from multiple sources and in some fashion interleaves their data streams into one satellite transmission stream (Figure 8.3b). As we saw earlier with multiple terminals sharing one multipoint line, the following two basic techniques exist for sharing the access to a satellite system:

■ **Frequency division multiple access (FDMA) with fixed assignment**: The total range of transmission frequencies assigned to the satellite is divided into **groups**, and each group of frequencies is assigned to a user of the satellite system. While this is the simpler technique, it has two major disadvantages: it is inflexible in that the assignment of frequencies to users does not change easily; and valuable frequencies are wasted as buffer regions to prevent the assigned channels from overlapping their frequencies.

■ **Time division multiple access (TDMA)**: The range of acceptable transmission frequencies is subdivided in time, allowing each user of the satellite system a unique time slot in which to transmit its data. Similar to statistical time division multiplexing introduced in Chapter 3, the assignment of time slots under TDMA operation is made on demand from the users.

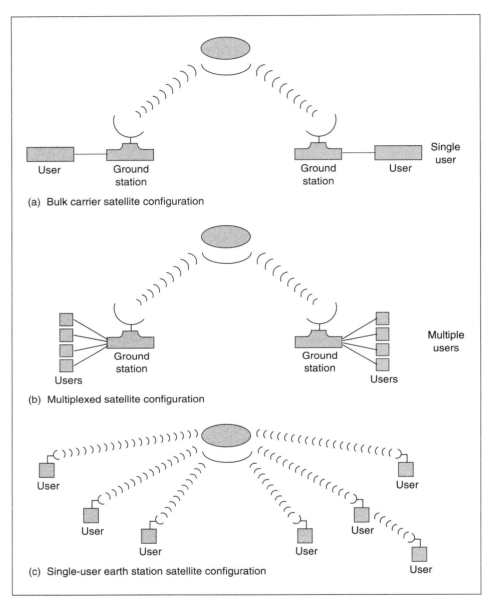

(a) Bulk carrier satellite configuration

(b) Multiplexed satellite configuration

(c) Single-user earth station satellite configuration

FIGURE 8.3 Three basic configurations of satellite systems

An interesting problem arises with time division multiple access. If the satellite transmission time slots are assigned on demand, how does a satellite system know which users wish to transmit data, and then how are the available time slots assigned to those users? The satellite system could poll each user to ask whether it has data to transmit, but so much time would be lost by the polling process that this technique would not be economically feasible. The satellite system could use a carrier-sense multiple access technique, but transmission distances are so long and transmission speeds are so fast that if a collision occurred between two users, they would not hear the collision until well after it happened. Too much time would be lost dealing with collisions and available time on the satellite system would be wasted.

An access method created for packet radio, Aloha, is another possible solution. Under the Aloha access method, a user station simply transmits without listening to the medium. If the user receives a response, the data must have gotten through. If no response is received, there must have been a collision, so the user has to try again. The Aloha method is clearly too wasteful of time for an expensive satellite system, and thus not an economic solution.

The one technique that seems to work best for access to a TDMA satellite system is a **reservation system**. With a reservation system, the users "place" reservations for future time slots. When a user-reserved time slot arrives, the user transmits its data to the satellite. Two basic types of reservation systems exist: centralized reservations, in which all reservations go to one central location and that site handles the incoming requests; and distributed reservations, in which no central site handles the reservations but the individual users come to some agreement as to the order of transmission.

The two techniques—frequency division multiple access and time division multiple access are not mutually exclusive. It is not uncommon to divide a satellite bandwidth into multiple frequency groups, then further divide a frequency group by time division multiplexing.

Single-User Earth Station

With a single-user earth station, each user of the satellite system has its own ground station and uses it to transmit its data to the satellite (Figure 8.3c). An example of a single-user earth station satellite system is the **Very Small Aperture Terminal** (**VSAT**) system. A VSAT user is its own ground station and has a small antenna (4 to 6 feet across). Among all the user ground stations is one master station that is typically connected to a mainframe computer system (Figure 8.4). The ground stations communicate with the mainframe computer via the satellite and the master station.

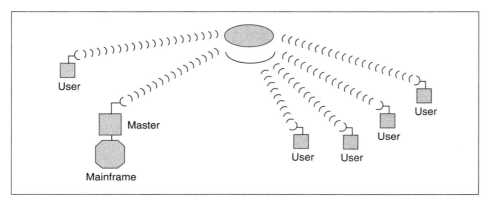

FIGURE 8.4   Configuration of a VSAT satellite system

Once again, the question arises: How do all the ground stations share the bandwidth of the one satellite? While TDMA and alternate forms of Aloha are used, the most popular solution is a reservation system where each ground station places its reservation on a separate channel.

RADIO NETWORKS

The next type of transmission system in our discussion is the radio network. Radio waves are an extremely popular form of transmission, both for analog and digital signals. AM radio, FM radio, and television are the most familiar uses of continuous, analog radio transmission. Police and fire departments use digital FM radio to transmit the necessary information to and from remote locations. When a police squad car pulls over another vehicle, the squad car radio set can send and receive detailed information concerning the apprehended vehicle. Pocket pagers that many people carry for business and personal uses receive digital FM signals that convey phone numbers or short messages. Even fast-food restaurants now use analog radio transmission, allowing their drive-through order-takers to use mobile headsets, thus allowing them the freedom to move about the restaurant.

A merger of radio systems and computer networks has led to the flexible **broadcast radio network**. Broadcast radio network systems involve two or more receiver/transmitters (ground stations) that communicate to each other by radio waves. Very often the communication is in the form of a packet of digitized data. A radio network has several characteristics that set it apart from a satellite system:

- All transmission is from ground station to ground station. No signals are bounced off orbiting satellites.
- Data is not transmitted at such high rates as with satellite systems. (Radio networks typically transmit thousands of bits per second; satellite networks transmit millions of bits per second.)
- While satellite transmissions are on a narrow beam, radio signals are omnidirectional (Figure 8.5).
- Radio network stations can easily be relocated.
- AM radio waves, because of their frequencies, can be aimed into space and bounced off the earth's ionosphere, allowing greater transmission distances than the line-of-sight transmission of FM radio waves.

One popular form of radio network is the **packet radio network**. This type of network transmits data between stations in the form of a packet, very similar to the frames created at the data link layer. Packet radio networks come in two basic forms: centralized and distributed. In a centralized system, all stations transmit data to one centrally located site, or master station, using a given set of frequencies (Figure 8.6). The master station responds to a station by transmitting a response on a second set of frequencies. Since all transmissions are omnidirectional, all stations receive transmissions from the master station. The maximum *radius* of a system is line of sight, or roughly 25 km. Thus, a station may not be more than 25 km from the master station.

The classic example of a centralized packet radio network is the Alohanet. Designed in Hawaii in the late 1960s, Alohanet was created to allow remote terminals wireless

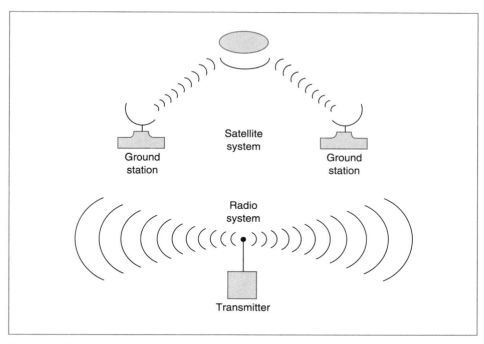

FIGURE 8.5   Satellite signals versus omnidirectional radio signals

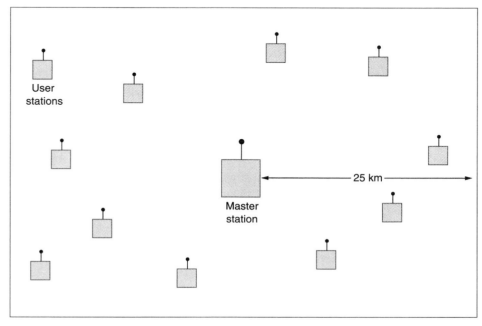

FIGURE 8.6   Centralized packet radio network

access to a mainframe computer. A terminal, wishing to transmit data to the mainframe, would package its data into a particular form and transmit it at 407 MHz. Data transmission speed was 9600 bps. A response from the mainframe would be transmitted at 413 MHz. Since all terminals transmitted at the same frequency, some sort of access protocol was necessary. The access protocol Aloha was created; it is discussed later in the chapter.

A more modern example of a centralized packet radio network is the type that allows rural area schools to communicate with a university for access to card catalogs and the Internet. Many rural schools want to interface their terminals and microcomputers to the mainframe computer at a regional university but cannot afford the costs they would incur using modems and long-distance telephone lines. A centralized packet radio network allows these schools and the university to exchange data in a more cost-efficient manner.

Distributed packet radio networks have no master station (Figure 8.7). All stations transmit to all other stations using one set of frequencies. Consequently, every station hears what any station transmits, like a local area network. Since all stations transmit to all other stations, the maximum *diameter* of the network is line of sight, or approximately 25 km.

AX.25 is a popular standard for distributed packet radio networks. This standard is very similar to HDLC, from which it has borrowed several features. Each packet transmitted begins with a flag field (01111110), followed by the address field. The address field can be from 14 to 70 octets in length, which should be sufficient for both sender and receiver addresses and any intervening repeater addresses. Following the address field is the control field (in HDLC format) and a protocol identifier, which specifies the type of network layer protocol used, if any. Next is the data field, a two-octet frame check sequence, and an ending flag (01111110).

Packet Radio Access Protocols

To allow multiple radio stations to transmit their data without collisions, some form of packet radio access protocol must be used. Three basic protocols exist: Aloha, slotted Aloha, and carrier sense multiple access.

Aloha

Created to work with the Alohanet, the Aloha protocol, or pure Aloha, is the simplest of the packet radio access protocols. A station wishing to transmit simply transmits. After a reasonable delay, the station should receive a response from the master station. If it does not, a collision is assumed, and the data packet is retransmitted. Since the station does not listen before talking, there is a nontrivial probability of collision. This increase in collisions affects the chances of a data packet getting through collision free. Studies have shown that typical throughput for an Aloha system is only 18%.

Slotted Aloha

To reduce the number of collisions in Aloha and increase the throughput, a variation of pure Aloha was created. Slotted Aloha organizes the time on the channel into uniform

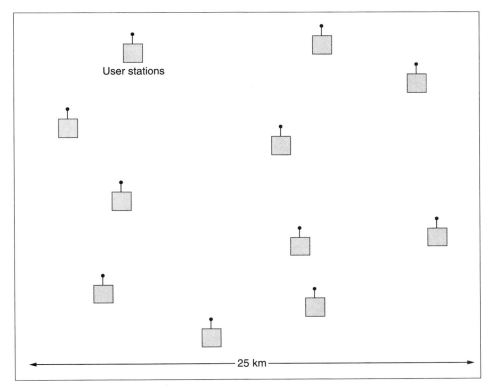

User stations

25 km

FIGURE 8.7   Distributed packet radio network

slots. Each slot size equals the maximum packet transmission time. All stations are synchronized so that a station may transmit only at the beginning of a time slot. In this way, collisions can occur only during a small time period. Slotted Aloha does reduce the number of collisions, and studies have shown that typical throughput for a slotted Aloha system is 37%.

CSMA (Carrier Sense Multiple Access)

Essentially the same as the local area network CSMA/CD but without the collision detection, CSMA works well when stations are relatively close together. This closeness decreases the signal propagation time between stations, thus reducing the collision window and the chance for a collision. Persistent, nonpersistent, and p-persistent CSMA protocols exist, with studies showing that the 0.01-persistent and nonpersistent protocols achieve 80–90% throughput.

CELLULAR PHONE SYSTEMS

Another transmission framework is the cellular phone system. Increasingly popular, this system is based on cellular radio. With cellular radio, the cellular radio market is

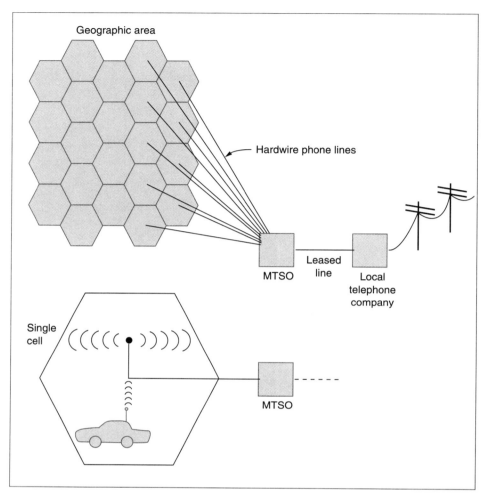

FIGURE 8.8   Cellular radio cells and the MTSO

broken into adjacent cells (Figure 8.8). Each cell contains a low-power transmitter/ receiver. A mobile phone within a cell communicates to the cell transmitter using frequency division multiplexing. The cell transmitter is connected to a mobile telephone switching office (MTSO) by hardwire telephone line. The MTSO is then connected to the local telephone company through a leased phone line, unless the MTSO and the local telephone company are one and the same. If the mobile phone moves from one cell to another, the MTSO hands off the connection from one cell to another.

Since each cell uses low-power transmissions, it is not likely that a transmission within a particular cell will interfere with the transmission in another cell that is more than one cell's distance away. Only adjacent cells need to use different sets of frequencies. Thus, the frequencies that are used within a cell can be reused in other cells, allowing for more phone connections than there are available channels.

In each cell at least one channel is responsible for the setup and control of calls. A mobile phone, wishing to place a call, attempts to seize a setup channel. When this is accomplished, the phone transmits its unique, built-in mobile ID number (MIN). This 10-digit (34-bit) MIN is transferred to the MTSO, where the MIN's account is checked for validity. If the phone bill has been paid, the MTSO assigns a channel to that connection, the mobile phone releases the setup channel, seizes the assigned channel, and proceeds to place the phone call.

While each mobile phone's mobile ID number is supposed to be unique (the 34-bit MIN is burned into a chip inside the phone), there is still room for cellular phone fraud. Unscrupulous users will monitor the air waves and tune into the frequencies used by cellular phone systems. While they listen electronically, they record the MINs that are transmitted for valid phone calls, and then fabricate new chips with these "stolen" MINs. Now they have a mobile phone that can identify itself with someone else's account and avoid paying for any cellular phone calls.

When someone tries to call a mobile phone, all cell base transmitters transmit the mobile phone's MIN (mobile ID number). When the MIN is recognized by the mobile phone, the mobile phone tries to seize the local cell's setup channel. When the setup channel is seized, the mobile phone sends its MIN to the MTSO, the MTSO verifies the MIN, a channel is assigned to the mobile phone, and the incoming call is connected.

At the present time, the FCC allows at most two companies to provide a cellular phone system within each cellular market. Thus, if you have a contract with company A to provide cellular service, and you travel to another city or state, you may find that the city you are visiting does not have services provided by your company (if services are provided at all!). Luckily, you are still able to place and receive calls using the local service, although special arrangements may be necessary, and you may incur additional costs.

Mobile cellular phones come in many varieties. Many people are familiar with the automobile cellular phone. There are also stand-alone units that people can carry with them. It is even possible to create a mobile office. In your vehicle you may have a cellular phone, a fax, and a printer all communicating over the cellular network. Some forecasters predict that the home telephone will become obsolete, replaced by the personal telephone that each person carries. Then, much like a popular television science-fiction show, a person traveling anywhere can call or receive phone calls from anyone else.

In the near future, the current frequency division multiplexing between mobile phone and cell base transmitter will probably be replaced with time division multiplexing. This transition should open the cellular market to many more possible subscribers. At the present time, it is not uncommon for your connection to be severed ("dropped") as your mobile phone moves from one cell to another. The problem is that too many mobile phones are being used in the area, and when you move into a new cell, there are no available channels for you to be handed off to.

TERRESTRIAL MICROWAVE

We have examined two types of radio-wave transmission—satellite systems and broadcast radio. Now we look at the third type: **terrestrial microwave**. Terrestrial microwave is similar to a satellite system in that a high-frequency narrow-beam radio signal is used to transmit data. The primary difference between terrestrial microwave and

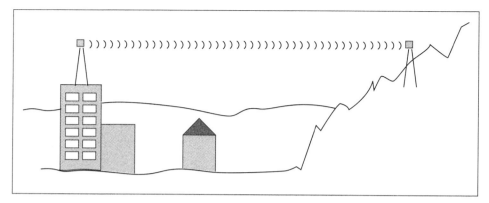

FIGURE 8.9   Terrestrial microwave relay stations on buildings, mountain tops, and towers

satellite is that while a satellite system transmits signals from earth to satellite back to earth, terrestrial microwave can use only line-of-sight transmission (Figure 8.9.). Given the placement of terrestrial microwave stations on top of mountains, high buildings, or towers, the typical line-of-sight distance is approximately 30 miles.

It is quite common to find a terrestrial microwave system used as a private transmission system between two or more user-owned facilities. While the costs of such systems are usually very high, and an FCC license is required before any transmissions occur, occasionally the demands outweigh the costs. Many times it is simpler and more economical to send your data to a third-party microwave transmission service, which, like a public data network service, will gladly transmit your data for you over an existing microwave network for a fee.

WIRELESS LOCAL AREA NETWORKS

A new breed of wireless LANs is beginning to emerge; it transmits data between workstations and a file server using high-frequency radio waves (902–928 MHz). Using an omnidirectional antenna, the workstation and file server may be separated by as much as 800 feet. It is also possible to incorporate a directional antenna, which then allows a distance of 8 kilometers (5 miles) between workstation and file server.

The network cards support the Ethernet protocol and simply consist of a single board inserted into an expansion-card slot within the workstation. Connected to the expansion board by means of a two-foot cable is the antenna, which can simply be affixed to the outside of the workstation (Figure 8.10). The hardware also allows the operation of a number of popular modern microcomputer network operating systems.

While simplicity of installation is clearly an advantage of a wireless local area network, a disadvantage is the relatively slow data transmission speed of 2 Mbps. However, for networks that do not perform major file transfers or support large numbers of simultaneous users, the wireless systems are enticing. With the rapid proliferation of radio transmission systems in the communications industry, it is a sure bet that the wireless local area network will grow as a viable alternative to wired local area networks.

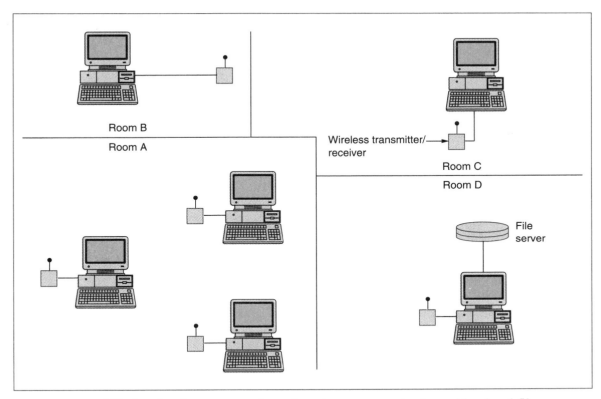

FIGURE 8.10   Wireless local area network workstations communicating with a local file server

CASE STUDY

To see the real-world application of wireless networks, we look now at their impact on our friends in the computing center. The XYZ University computing center is starting to grow weary of installing or upgrading new cabling every time someone desires a new or improved service. Since wireless transmission systems are growing more popular every day, perhaps the computing center could eliminate all or most of the physical cabling and produce a wireless campus. Let's examine each of the transmission techniques presented and determine whether one of them is a feasible alternative to physical cables.

Since satellite systems were introduced first, we will consider them first. Satellite systems are designed for applications that transfer high amounts of data over long distances. The XYZ University campus is not even one kilometer wide, thus eliminating the need for data transfer over long distances, a major advantage of a satellite system. Also, the computer staff do not expect a great amount of data traffic between buildings. Individual local area networks may see a certain amount of data traffic between workstations and local servers, but this would not merit the expense of a satellite.

Terrestrial microwave might also be a poor choice for the XYZ campus. While more reasonable than a satellite system, terrestrial microwave has the disadvantages

The labels in the figure read: Room B, Room A, Wireless transmitter/receiver, Room C, Room D, File server.

of high costs and the necessity of FCC licensing. The computing center staff could install and operate their own system, but they can't justify it.

Broadcast radio, the least expensive technique thus far, has interesting possibilities. A centralized packet radio network could be created with the computing center as the central hub. Each building could have a network transmitter that communicates with the hub. An unfortunate disadvantage of a broadcast packet radio network is the low data transfer speeds, typically 9600 bps.

Perhaps the most exciting possibility is the wireless local area network. Each building could contain multiple wireless workstations that would communicate with a wireless network server located in the computing center. This scenario, however, might lead to serious congestion problems if all workstations communicated with one server. A more likely solution would be to allow each building one or more separate wireless local area networks, each with its own network server. Each network could then communicate to the computing center using a wireless bridging device.

A possible disadvantage is once again the slow transmission speed of wireless local area networks. Perhaps the campus does not need to move toward a completely wireless operation, but could maintain a high-speed physical backbone interconnecting each building and the wireless local area networks. While the backbone would be a high-speed transfer device, the wireless networks would still be limited to their slower maximum data transfer speeds.

Having considered the various wireless techniques, the computing center has decided to compromise. Rather than creating an entirely wireless campus, the staff has decided to install a wireless local area network for the next department that submits a request. Usage of the wireless network will be monitored and further consideration will be given to future expansion into a wireless campus.

SUMMARY

Satellite and radio broadcast networks are an increasingly popular form of communication in the world of data communications and computer networks. Among the many examples of satellite and radio broadcast networks are: long-distance television transmission, cellular phones, facsimile, wireless message transmission (pagers), telegraphy, closed caption television (Teletext), burglar alarm systems for home and automobile, emergency radio systems, citizens band radio systems, and global positioning systems. Several advantages set these networks apart from other types of networks, such as high data transmission rates and ability to transmit over long distances.

A satellite communication system is essentially two microwave transmission systems: the uplink from ground station to satellite, and the downlink from satellite back to ground station. Satellite systems can be configured into three basic topologies: bulk carrier facilities, multiplexed satellite configuration, and single-user earth station configuration. Access to the satellite's system in a multiplexed configuration is accomplished by either frequency division multiple access or time division multiple access with reservation.

A merger of radio systems and computer networks has led to the flexible radio network. Radio network systems involve two or more receiver/transmitters (ground stations) that communicate to each other by radio waves. One popular form of radio network is the packet radio network. A packet radio network transmits data between stations in the form of a packet, very similar to the frames created at the data link layer. Packet radio networks come in two basic forms: centralized and distributed. A modern example of a centralized packet radio network is the type that allows rural schools to communicate with a university for access to card catalogs and the Internet. The AX.25 is a popular standard for distributed packet radio networks. There are three basic packet radio access methods: Aloha, slotted Aloha, and carrier-sense multiple access.

The popular cellular phone systems are based on cellular radio, in which a mobile cellular phone communicates with a cell transmitter by frequency division multiplexing. Each cell transmitter is connected to the mobile telephone switching office, which is further connected to the local telephone service. The growing demand for cellular phone systems is leading to additional frequencies and time division multiplexing techniques.

EXERCISES

1. What are the advantages that set broadcast and radio networks apart from other networks?
2. List the three basic topologies of satellite systems with a sentence describing each.
3. How do frequency division multiple access and time division multiple access compare to frequency division multiplexing and time division multiplexing, described in Chapter 4?
4. Sketch a rough drawing showing how carrier-sense multiple access is an ineffective access method for two ground stations and one satellite.
5. List the similarities between AX.25 and HDLC.
6. List the three packet radio access methods with their relative throughput values.
7. A cellular phone handset is considered to be essentially impervious to theft. Explain how this is possible.
8. Draw a cellular phone market with 18 adjacent cells. Assign a frequency range to each cell. How many different frequency ranges are necessary to ensure that no two adjacent cells share the same frequency range?
9. Why would someone using a cellular phone in his or her automobile and driving across a large metropolitan area all of a sudden lose the phone connection? (It has nothing to do with paying the bill.)
10. List the similarities and differences between a satellite system and a terrestrial microwave system.
11. Some networks, such as wireless local area networks, employ a slight variation of the CSMA/CD protocol, termed CSMA/CA (carrier-sense multiple access with collision avoidance). Speculate how CSMA/CA might differ from CSMA/CD. What would be the advantage of using CSMA/CA with a wireless local area network?

CHAPTER 9

TRANSPORT AND SESSION LAYERS

INTRODUCTION

To this point, we have covered the first three layers of the OSI model: the physical layer, the data link layer, and the network layer. We can consider these point-to-point layers, as the emphasis has been on passing the data from one point to the next within a network. Figure 9.1 demonstrates a simple network with endpoints A and B and three intermediate points X, Y, and Z (which could be bridges, repeaters, or computer nodes). The physical, data link, and network layers are concerned with the traffic flow between A and X, between X and Y, between Y and Z, and between Z and B.

The upper layers—transport, session, presentation, and application—are concerned with the flow of information between the two endpoints, A and B. Thus, the upper layers are often termed the *end-to-end* layers.

Most data communications study addresses primarily the lower three layers, but modern computer networks and their applications cannot function without the benefit of the transport layer and its higher cousins. The transport layer is the first of these upper layers and communicates directly with the network layer, so we examine it first. We will see how important the transport layer is; it has the responsibility of providing an error-free end-to-end connection regardless of the type of underlying network. This is no simple task. After discussing the transport layer, we examine the simpler session layer and its responsibilities.

TRANSPORT PROTOCOL CLASSES

The transport layer's primary responsibility is to provide an error-free end-to-end connection given any type of underlying network. Since there are so many different types

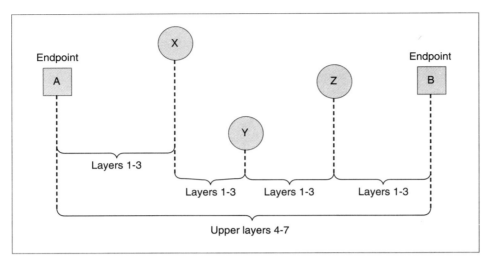

FIGURE 9.1   Physical, data link, and network layer traffic flow between points within a network

of networks, this is no small feat. Thus far we have examined public data networks, the Internet, radio networks, satellite networks, and several types of local area networks. A transport protocol that can provide an error-free connection over all these types of networks would indeed be enormous. Fortunately, the designers of the OSI model tried to make things a little bit simpler by creating a set of **transport protocol classes**. The choice of a transport protocol class is then determined by the type of the underlying network.

ISO (International Standards Organization) groups networks into the following three types:

- **Network Type A**: The Type A network is nearly flawless, with error-free packet delivery service and no network resets. If a packet is lost or garbled, the data link or network layer recognizes this and corrects the situation. A network reset, in which the network experiences difficulties and transmits some form of reset condition, is rare. Networks of Type A are indeed few and far between. Local area networks with their extremely low error rates often qualify as Type A networks.

- **Network Type B**: With Type B networks, packet delivery service is nearly flawless, as in Type A networks, but the network may occasionally issue network resets. If a network issues a reset and backs up to a previous point, the transport layer may begin receiving data packets it has received before, or may receive packets out of order.

- **Network Type C**: A Type C network is an unreliable network, with lost or duplicated data packets and occasional network resets. This is the worst kind of network, and surprisingly, the most common. Essentially, the network is saying, "I will get the data there if I can, when I can, with no guarantee of delivery." A Type C network requires the most rigorous transport layer protocol. If the network is not going to guarantee error-free delivery, something must, and that something is the transport layer.

Operating over these three types of networks are transport protocols. The following five classes of transport protocols—from 0 to 4—have been defined:

- **Transport Protocol Class 0: Simple Class**: This is the simplest class of transport protocol and operates over Type A networks. Since the Type A network is essentially error free with no network resets, the Class 0 transport protocol has a pretty easy job. Most of the error checking has already been done by the data link and network layers, leaving little responsibility for the transport layer.

- **Transport Protocol Class 1: Basic Error Recovery Class**: This class of transport protocol operates over Type B networks. Since Type B networks may experience a number of network resets, the transport layer is prepared to react to these by keeping both endpoints of the connection synchronized. It accomplishes this synchronization by inserting in each transport data packet a sequence number. Both endpoints of the transport layer record the sequence number of each data packet and take the necessary action when a packet arrives with an out-of-sequence number.

- **Transport Protocol Class 2: Multiplexing Class**: Class 2 protocols are designed to work over Type A reliable networks, but unlike Class 0 protocols, Class 2 protocols can multiplex multiple transport connections over the same network connection. Thus, there may be multiple end-to-end connections sharing the same path through a network.

- **Transport Protocol Class 3: Error Recovery and Multiplexing Class**: Class 3 protocols operate over Type B networks and combine the basic error recovery techniques of Class 1 protocols with the multiplexing techniques of Class 2 protocols.

- **Transport Protocol Class 4: Error Detection and Recovery Class**: Class 4 protocols are designed for Type C unreliable networks. The transport protocol assumes the worst can happen at the network layer; it handles garbled, lost, or duplicate data packets and can recover from network resets. This transport protocol also performs transport connection multiplexing.

The choice of which transport protocol class to use is determined when the two endpoints of a "connection-to-be" begin the negotiations for a connection, or during the connection establishment time. If the two endpoints are using an unreliable network service and wish to transmit data in a reliable fashion, they select the appropriate transport protocol class. If the two ends of the connection cannot agree on an acceptable transport protocol class, a lesser transport connection is negotiated. If the two endpoints cannot agree on a lesser transport connection, the connection is dropped completely.

For example, when one endpoint wishes to establish a transport connection, it transmits a connection establishment packet containing a number of **Quality of Service (QOS)** parameters. These parameters specify *preferred*, *acceptable*, and *unacceptable* values for a number of events, such as transit delay times, residual error rates, and throughput rates. When the receiving endpoint receives the connection establishment packet with the QOS parameters, that endpoint determines whether it can provide a transport connection using the *preferred* values of the QOS parameters. If it cannot, but a connection can be established using the *acceptable* values, a connection is established using the lesser values. If the receiving endpoint cannot provide a connection using even the acceptable values, the original endpoint is informed that no connection at all can be established.

Basic Transport Layer Responsibilities

The previous section showed that five classes of transport protocols have been defined to produce an error-free end-to-end network. To create this error-free network, the transport layer has several basic responsibilities. By performing some or all of these basic responsibilities, the five classes of transport protocols may achieve their common goal: an error-free connection. While it is possible to list over 20 separate responsibilities of the transport layer, we will examine only the eight most common ones: (1) establishing a connection, (2) releasing a connection, (3) flow control, (4) multiplexing, (5) crash recovery, (6) data transfer, (7) splitting of long packets, and (8) expedited delivery.

Establishing a Connection

The physical, data link, and network layers handle the details from node to node within the network; the transport layer is the first layer encountered that handles the end-to-end connections. It is responsible for establishing the end-to-end network connection. While the end-to-end connection can seem as simple as endpoint A sending a message to endpoint B, saying, "Let's establish a connection," and endpoint B responding, "OK, let's do it," it is really much more involved. To begin, how does endpoint A know the address of endpoint B? If a user at location A wishes to perform a set of queries from a database at location B, how does user A know the address of the database server? A device called a **name server** could be called by endpoint A, which would provide the address of the requested service at endpoint B. So that everybody knows what the address of the name server is, the address is given to all users before they begin requesting connections.

A second issue that merits attention is connection acknowledgment. If endpoint A asks endpoint B to establish a connection, and endpoint B agrees with an acknowledgment, how does endpoint B know that endpoint A received the acknowledgment? So that both endpoints understand that a connection has been agreed on, a **three-way handshake** is used. Endpoint A asks for the connection; endpoint B acknowledges; and endpoint A acknowledges the acknowledgment. (Of course, you could ask how endpoint A knows that endpoint B received the acknowledgment of the acknowledgment, but this could go on forever. A three-way handshake will have to be sufficient.)

A third point to consider is connection negotiation. Both endpoints must agree during connection establishment on which transport protocol class to use. If one endpoint cannot perform the requested class of transport protocol, both endpoints must either agree to a lower class of protocol or agree that no connection at all can be made. The transport protocol must be constructed so that the two endpoints can perform this connection negotiation.

Releasing a Connection

When one endpoint or the other wishes to terminate or release a connection, how is the other endpoint informed of this requested release? For example, endpoint A may have finished sending data to endpoint B and may request a termination of the connection, but endpoint B may have more data to send to A, and thus want to continue the connection. So both endpoints agree that a connection termination is desired, a three-way handshake is also needed for releasing a connection. The first endpoint states it wishes to terminate the connection, the second endpoint agrees, and the first endpoint acknowledges the second endpoint's agreement.

Flow Control

We have stated numerous times that each layer of the OSI model has two basic responsibilities. The first is to accept an outgoing data packet from the next higher layer, perform some or no services for the data packet, and pass the updated data packet to the next lower layer. In performing these services, the layer may add additional fields of information (encapsulated) to the data packet received from the upper layer. The second basic responsibility of each layer is essentially the reverse of the first. Each layer accepts an incoming data packet from the next lower layer, removes any fields added by its corresponding layer on the transmitting side of the connection, and passes the data packet on to the next higher layer (Figure 9.2).

Operating in the same fashion, the transport layer accepts outgoing data packets from the session layer, performs any required services, adds the necessary header information, and passes these packets to the network layer. It also accepts incoming data

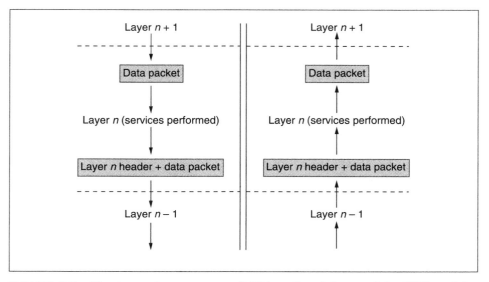

FIGURE 9.2   The two primary responsibilities of each layer of the OSI model

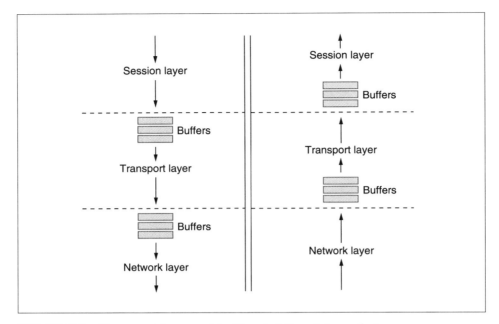

FIGURE 9.3   Transport layer and buffers holding data packets

packets from the network layer, removes any transport layer headers, and passes these data packets to the session layer. To handle all this giving and taking of data, buffers are necessary to keep things moving smoothly (Figure 9.3).

So that the buffers do not overflow, some form of flow control is needed between the two endpoints. This way, if endpoint A is becoming overwhelmed with data arriving from endpoint B, endpoint A can tell endpoint B to slow down. This function is similar to the flow control we saw at the data link layer. The difference between transport layer flow control and data link layer flow control is that data link flow control affects the flow between two nodes in the network, while transport flow control affects the flow between the two endpoints of the network connection.

Multiplexing

To use a network connection more efficiently, you should be able to multiplex multiple transport connections over a single network connection. This could be advantageous if network connections are expensive or difficult to obtain. When several transport connections share one costly network connection, it is possible for multiple slower transport connections all to travel over the same higher-speed network connection. This is an example of transport layer **upward multiplexing**.

It is also possible for a single transport connection to use multiple network connections. This would allow an extremely busy transport link to send data over multiple slower network connections. This is an example of **downward multiplexing**. Figure 9.4 demonstrates upward and downward multiplexing between the network and transport layers.

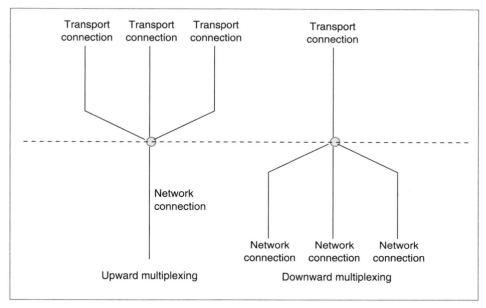

FIGURE 9.4   Upward and downward multiplexing

Crash Recovery

Each endpoint of the transport layer should buffer one or more transmitted data packets; with this protection, if the underlying network experiences a problem, both endpoints of the transport layer can back up to a mutually agreeable safe point and resume transmission. A simple numbering of the transport data packets sent and received may be sufficient to allow both endpoints to back up to a known packet and recover from network problems.

Data Transfer

Each transport protocol data packet will contain a data field that allows the transfer of data between the two endpoints. This data field is the data packet that has been passed down from the next higher layer in the application. All five transport protocol classes provide for this feature.

Splitting of Long Packets

One of the responsibilities of the transport layer is to transmit data packets from one endpoint to the other using the services of the underlying network. What happens, however, if the transport layer wishes to transmit a packet of size n over a network that is capable of transmitting only packets of a smaller size m? It would be desirable to split long transport data packets into sizes that will fit into the data packet size of the underlying network. If long packets are split into multiple shorter packets, the transport layer must somehow recognize the multiple shorter segments so they can be reassembled into the original packet at a later time.

Expedited Delivery

So far, we have always assumed that before any data is exchanged between endpoints, a connection between the two endpoints has been established. But what if the user at endpoint A wishes to send only a single packet of data to a user at endpoint B? Or a connection has been established between two endpoints, but there is a need to send an immediate data packet out of sequence from the previous data packets sent? An example of this might be sending an error message, or sending a break command to stop the inbound flow of data packets. What is needed is **expedited data**. This is a unique data packet that may not include flow control sequence numbers and immediately passes through any buffers to arrive at its destination as quickly as possible.

Table 9.1 summarizes the eight transport layer responsibilities and shows which transport protocol class performs which responsibility.

SAMPLE TRANSPORT LAYER PROTOCOL: TCP

Perhaps one of the most common examples of a transport layer protocol is the second half of the popular TCP/IP standard: **TCP (Transmission Control Protocol)**. As we discovered in Chapter 7, TCP/IP is the Department of Defense's standard for inter-networking heterogeneous (dissimilar) networks. The function of TCP is to turn an unreliable subnetwork into a reliable network, free from lost and duplicate packets. To accomplish this, TCP places a header on the front of every data packet that travels from one endpoint to the other. This header contains several fields and allows TCP to accomplish many of the eight responsibilities introduced in the previous section.

TABLE 9.1   Summary of Eight Transport Layer Responsibilities

Responsibility	Transport Protocol Class				
	0	1	2	3	4
Establishing a connection	Y	Y	Y	Y	Y
Releasing a connection	Y	Y	Y	Y	Y
Data transfer	Y	Y	Y	Y	Y
Splitting of long messages	Y	Y	Y	Y	Y
Flow control			opt	Y	Y
Resynchronizing after network reset		Y		Y	Y
Expedited data service		opt	opt	Y	Y
Multiplexing			Y	Y	Y

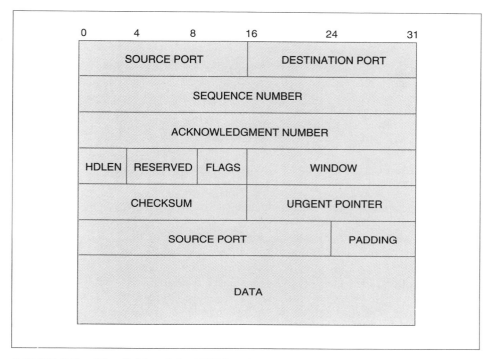

FIGURE 9.5   The fields of the TCP header

The best way to explain the operation of TCP and at the same time examine which of the eight responsibilities it satisfies is to examine the TCP Header field by field. Figure 9.5 shows the fields of the header.

The first two fields, Source Port and Destination Port, contain the addresses of the application programs at both ends of the transport connection. Each address is fixed at 16 bits and represents some form of address at the transport layer. The Sequence Number field contains a 32-bit value. This appears to be a very large sequence number if one believes the sequence number counts TCP packets. It doesn't. The sequence number counts bytes and indicates the data position of this packet relative to the entire stream of data. The next field, Acknowledgment Number, indicates the number of the byte of data that the source expects to receive next. The HDLEN field is a 4-bit integer that indicates how many 32-bit words are contained within the TCP Header. Since the last field, the Options field, can be of variable length, the HDLEN field is necessary to mark the end of the TCP header and the beginning of the data portion of the packet.

The Flag field contains six flags that are used to convey the intent of the TCP packet (Figure 9.6). The URG field is a single bit that indicates whether the following TCP packet contains urgent, or expedited, data. The ACK and SYN fields combine to denote a connection request by one endpoint and a connection confirm by the second endpoint. The PSH bit indicates that this particular transport packet is requesting a push and the receiving station is not to wait for its incoming buffer to fill before processing. The RST bit is used to perform a reset connection in the event that something happens to the underlying network and the two transport endpoints must resynchronize. The FIN

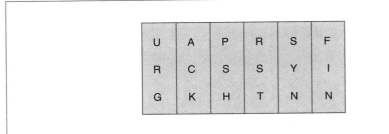

FIGURE 9.6   The six flags in the FLAG field

bit is used by the sending endpoint to indicate that it has no more data to transmit in this byte stream.

Returning to Figure 9.5, the Window field contains the sliding window value that provides flow control between the two endpoints. It is a 16-bit field because once again the unit counted is bytes, not packets. The Checksum provides for a 16-bit arithmetic checksum of the data field that follows the header. The Urgent Pointer is used in conjunction with the URG bit and is a byte offset from the current Sequence Number at which the urgent data may be found. The final field of the TCP Header is the Options field, which contains optional information such as connection establishment parameters.

In summary, TCP performs all the eight basic responsibilities of a transport layer. TCP allows for the creation of a transport session by using the ACK and SYN bits to denote connection request and connection confirmation. Each TCP packet contains room for data being transferred, and large data packets may be broken into smaller data packets using the Sequence field to denote the continuation of data from a previous packet and the FIN bit to denote the end of the stream. Flow control is accomplished with the Window field, and the RST bit can be used to indicate a reset condition in the connection. Expedited or urgent data is possible with the use of the URG bit and the corresponding Urgent Pointer field. Finally, it is possible to multiplex multiple transport connections. By selecting unique Source Port and Destination Port addresses, multiple unique transport connections may be established at the same time, over the same network connection.

THE SESSION LAYER

The session layer, which resides immediately above the transport layer, is often considered the forgotten layer. In fact, it is common to find a network application that follows the OSI model but does not contain a session layer. Nonetheless, the session layer can perform a number of vital services. Three of these are dialogue management, synchronization, and activity management. Next, we examine each of these and end with a topic related to the session layer, the remote procedure call.

Dialogue Management

Typically, when a network application uses the upper layers of the OSI model, a full-duplex connection is established. Sometimes, however, it may be necessary to create a half-duplex connection and allow the session layer to provide **dialogue management**. As an example, many types of database applications need only a half-duplex connection. The user of the database performs a data enquiry; the database responds with the requested data. There is no need for simultaneous two-way transfer of data. If such a half-duplex connection is desired, the session layer can provide the means. When two entities have negotiated for a half-duplex connection, an imaginary token is created. Only the entity holding the token may transmit data. So that both ends of the connection may participate in data transfer, it is possible to issue a session command (*S-Token-Give*) to move the token to the other entity. An additional session command (*S-Token-Please*) allows the entity not holding the token to ask the entity holding it to pass the token.

Synchronization

The second major function the session layer can perform is **synchronization** between the two session applications in case an error condition arises and the two endpoints have to drop back to a known point. Many times a network session is created to transmit large amounts of data, such as a book, from one point to another. If both endpoints agree during session creation, it is possible to insert a **sync point** at the end of each chapter. Assume that such sync points have been agreed on and we are currently in the middle of transmitting chapter two when a network failure occurs. After the network problem has been remedied, both ends of the session connection can agree to back up to the sync point at the end of chapter one and restart from that point. As it turns out, there are actually two types of sync points: **major sync points**, and **minor sync points**. Minor sync points might be inserted at the ends of pages of the book, while the major sync points could be inserted at the ends of the chapters. Since the transport layer can also provide a method of synchronization and backup, synchronization at the session layer is often unnecessary.

Activity Management

The third major function of the session layer is **activity management**. Activity management involves sending special messages at the beginning and end of an activity. The nature of the activity is defined by the session applications at either endpoint. As an example, if one user wishes to transfer multiple files to another user, the source user can insert an *S-Activity-Start* command at the beginning of each file, and an *S-Activity-End* command at the end of each file.

Another use of activity messages is to provide a **quarantine service**. This service, when agreed on by both endpoints of a session connection, requires the receiver of the data not to process any received data packets until an *S-Activity-End* command has been received. This forces the receiver to accept all necessary data packets before processing any. An example might be found in the banking industry when account transactions are transmitted to or from a remote site.

REMOTE PROCEDURE CALL

Chapter 5, "The Network Layer," introduced the concepts of connection-oriented and connectionless networks. The OSI model was based on a connection-oriented strategy.

However, many networks are connectionless. If an application wishes to request information from a service, it could spend the time required to negotiate a connection-oriented session, but this does not seem practical if the application is to perform a single quick function. For example, an application that performs a single query on a database server might not need to negotiate a session if there is a procedure that can produce results similar to those achieved with a session connection but remain connectionless. The **remote procedure call** is one such procedure. Not currently a procedure under the OSI model, the remote procedure call is an example of the increasingly popular **client-server model**. In the client-server model, a client performs a request for some operation. The operation could be a database query, a file server transfer, or any operation that one computer could call upon a second computer to perform. The server, after receiving the request, performs the operation and returns the requested information (Figure 9.7).

The remote procedure call allows a client application to request a service from a server by emulating the call of a remote procedure. Thus, similar to one local procedure calling a second local procedure with parameter passing, a client procedure calls a remote server procedure passing the appropriate parameters. The actual operation is, of course, a little more complicated. Both the client and server possess a library routine called a **stub** that handles the details of the actual data transfer. Thus, the client calls its stub passing parameters as in a normal procedure call. The stub takes the parameters and packs them into a message (parameter marshaling), which is given to the transport layer. The transport layer passes the message to the network and the message is transmitted to the server's transport layer. This layer then passes the message to the server stub, which unmarshals the parameters and gives them to the server. The server performs the requested operation and returns the results in the same manner that the request was transmitted. Figure 9.8 demonstrates a sample client-server interaction.

While the overall operation seems fairly straightforward, there are a few wrinkles that complicate the procedure. Any pass-by-reference parameters transmitted from client to server will create a problem because the client and server use two separate address spaces and a pass-by-reference address will have no local meaning. To get around this, some systems warn, "Do not use pass-by-reference parameters." This may

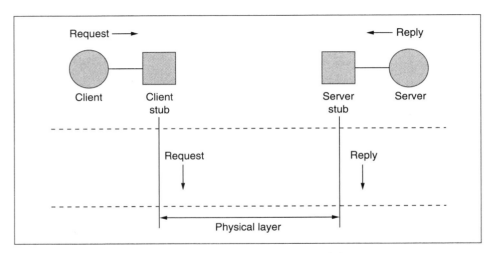

FIGURE 9.7   Simplified diagram of client-server model

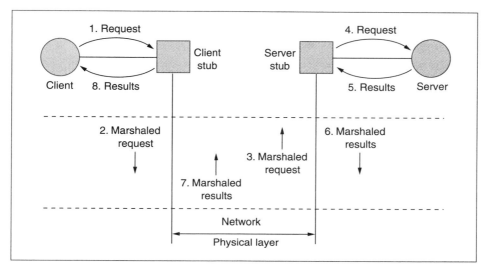

FIGURE 9.8   Client-server interaction

be necessary, but a goal of the remote procedure call was transparency, and if you force the client to take special precautions, it is no longer transparent.

A second problem that arises is a problem we have seen before: How does the client stub know whom to call? One solution that has been suggested requires all server stubs to inform a third party of their addresses so that when client stubs need to call a server, they can check first with this third party for the proper address.

There is a third problem: What happens if the server crashes after a client issues a remote procedure call but before the server has a chance to issue a response? Should the client stub wait forever for a response from a crashed routine? Should the client stub time out and report an error condition to the client? Should the client stub take matters into its own hands and retransmit the request? The last option could lead to duplicate requests, which could prove interesting if the request is to transfer money from one account to another.

The last problem we will consider is this: What happens if the client crashes after it has started a remote procedure call? This is called an **orphan condition** and can lead to locked files, wasted resources, and duplication of requests. Much research has been expended on this topic and several working solutions have been suggested.

CASE STUDY

Let's return now to the university computing center to see how the staff there can use the new information on transport and session layers. The XYZ University computing center established a connection between the XYZ campus and the Internet in Chapter 7. The primary software used to perform the interconnection is based on the TCP/IP standards from the Department of Defense. At the time they made the Internet connection, the computing center staff understood only the IP portion of the software. Having read Chapter 9, they now have a better understanding of the TCP portion of the software.

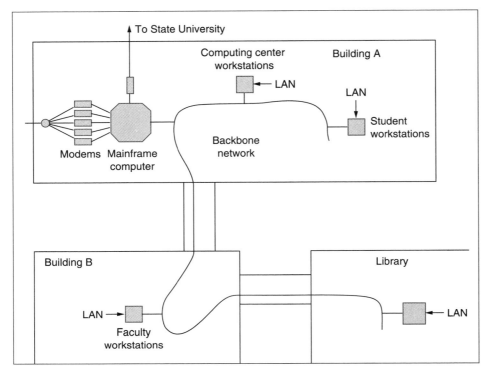

FIGURE 9.9 XYZ University campus computer interconnections

The interconnection of computer workstations for the XYZ campus as it currently exists is shown in Figure 9.9. The computing center staff wants to follow a single data packet from its creation in Building B until it reaches the Internet at State University. Following a single packet will allow the staff to monitor more closely and understand better what a data packet must go through as it progresses across the various campus networks.

We begin at a workstation in Building B, where a person using electronic mail enters the message "Meeting at 10:00 a.m." Since this message is bound for someone in another part of the country, the Internet's mail system will be used. Thus, the message must have a TCP header attached to the front of it:

SOURCE PORT   DEST PORT   SEQ NUM   ACK NUM   HDLEN   RES
FLAGS   WINDOW   CHECK   URG PTR   PADDING   "Meeting at
10:00 a.m."

This data packet with TCP header is then passed to the IP layer, which places its header on the front:

VERS   HLEN   SERVICE TYPE   TOT LEN   IDENT   FLAGS
FRAGMENT OFFSET   TIME TO LIVE   PROTOCOL   HEADER CHECK
SOURCE IP ADDR   DEST IP   ADDR   PADDING   **SOURCE PORT**

DEST PORT   SEQ NUM   ACK NUM   HDLEN   RES   FLAGS
WINDOW   CHECK   URG PTR   PADDING   "Meeting at 10:00 a.m."

Next, this TCP/IP data packet is to traverse the token ring network within Building B with an intermediate destination of the bridge between the Building B local area network and the campus backbone CSMA/CD network. Thus, the TCP/IP packet is passed to the LLC sublayer of the token ring network, which adds two LLC sublayer service access point addresses (DSAP and SSAP) and a control field:

DSAP SSAP CONTROL   VERS HLEN SERVICE TYPE TOT LEN
IDENT   FLAGS   FRAGMENT OFFSET   TIME TO LIVE   PROTOCOL
HEADER   CHECK   SOURCE IP ADDR   DEST IP ADDR   PADDING
SOURCE PORT   DEST PORT   SEQ NUM   ACK NUM   HDLEN   RES
FLAGS   WINDOW   CHECK   URG PTR   PADDING   "Meeting at
10:00 a.m."

The LLC sublayer then passes the data packet to the MAC sublayer for the token ring, which adds five fields to the front of the packet and appends three fields on the end of the packet:

START DELIM   ACCESS CONTROL   FRAME CONTROL   DEST ADDR
SOURCE ADDR   DSAP   SSAP   CONTROL   VERS   HLEN
SERVICE TYPE   TOT LEN   IDENT   FLAGS   FRAGMENT OFFSET
TIME TO LIVE   PROTOCOL   HEADER CHECK   SOURCE IP ADDR
DEST IP ADDR   PADDING   SOURCE PORT   DEST PORT   SEQ NUM
ACK NUM   HDLEN   RES   FLAGS   WINDOW   CHECK   URG PTR
PADDING   "Meeting at 10:00 a.m."   FRAME CHECK SEQ   END
DELIM   FRAME STATUS

This MAC sublayer packet is finally given to the physical layer of the token ring and is transmitted to the bridge between the token ring network and the CSMA/CD campus backbone network. Since the two networks are not the same, the bridge must convert the token ring MAC sublayer packet into a CSMA/CD sublayer packet. Thus, the bridge strips off the following fields:

START DELIM   ACCESS CONTROL   FRAME CONTROL   DEST ADDR
SOURCE ADDR

and

FRAME CHECK SEQ   END DELIM   FRAME STATUS

leaving:

DSAP   SSAP   CONTROL   VERS   HLEN   SERVICE TYPE   TOT LEN
IDENT   FLAGS   FRAGMENT OFFSET   TIME TO LIVE   PROTOCOL
HEADER   CHECK   SOURCE IP ADDR   DEST IP ADDR   PADDING
SOURCE PORT   DEST PORT   SEQ NUM   ACK NUM   HDLEN   RES
FLAGS   WINDOW CHECK   URG PTR   PADDING   "Meeting at 10:00 a.m."

and inserts the appropriate CSMA/CD MAC sublayer fields:

PREAMBLE   START FRAME DELIM   DEST ADDR   SOURCE ADDR
LENGTH   DSAP   SSAP   CONTROL   VERS   HLEN   SERVICE TYPE
TOT LEN   IDENT   FLAGS   FRAGMENT OFFSET   TIME TO LIVE
PROTOCOL   HEADER   CHECK   SOURCE IP ADDR   DEST IP ADDR
PADDING   SOURCE PORT   DEST PORT   SEQ NUM   ACK NUM
HDLEN   RES   FLAGS   WINDOW CHECK   URG PTR   PADDING
"Meeting at 10:00 a.m."   PAD   FRAME CHECK   SEQ

The CSMA/CD data packet traverses the CSMA/CD backbone and arrives at the computing center bridge, where all the CSMA/CD fields and LLC sublayer fields are removed, leaving only the TCP/IP packet:

VERS   HLEN   SERVICE TYPE   TOT LEN   IDENT   FLAGS
FRAGMENT   OFFSET   TIME TO LIVE   PROTOCOL   HEADER CHECK
SOURCE IP ADDR   DEST IP ADDR   PADDING   SOURCE PORT   DEST
PORT   SEQ NUM   ACK   NUM   HDLEN   RES   FLAGS   WINDOW
CHECK   URG PTR   PADDING   "Meeting at 10:00 a.m."

The computing center bridge recognizes this packet as one destined for the Internet. To get to the Internet connection at State University, the TCP/IP packet must first traverse the link between the two universities. If this link is a public data network (PDN), the TCP/IP packet must be converted into a PDN packet using the X.25 standard. If the link is not a PDN link, then the TCP/IP packet is modulated into the appropriate form for transmission.

Finally, the TCP/IP packet arrives at State University, its destination address is examined, the router at State University determines that the message is destined for another state, and the TCP/IP packet is sent on its way over the Internet.

While we have purposely omitted many details, you can see that the staff of the computing center is beginning to understand the amount of work that must be done for a single message to reach its destination. At numerous points along its journey, the original message could encounter checksum errors, wrong addresses, high-priority tokens, busy CSMA buses, overflowing buffers, inoperative telephone links, "time-to-live" violations, "hop-count" violations, and poorly written software. The staff has a new appreciation for what at first appears to be a simple thing.

SUMMARY

The upper layers—transport, session, presentation, and application—are concerned with the flow of information between the two endpoints. Although most study of data communications concentrates on the lower three layers, modern computer networks and their applications cannot function without the transport layer and its higher cousins.

The transport layer's primary responsibility is to provide an error-free end-to-end connection regardless of the underlying network. Five classes of transport protocols have been defined: simple class, basic error recovery class, multiplexing class, error recovery and multiplexing class, and error detection and recovery class. Whichever transport class is chosen, the transport layer can perform eight basic responsibilities:

establishing a connection, releasing a connection, flow control, multiplexing, crash recovery, expedited delivery, data transfer, and splitting of long messages. Perhaps one of the most common examples of a transport layer protocol is the popular TCP standard (Transmission Control Protocol).

The session layer, which is next up after the transport layer, is often the least used one of the OSI model. Nonetheless, the session layer can perform a number of vital services. The first service creates a half-duplex connection and allows the session layer to provide dialogue management. The second service the session layer can perform is synchronization between the two session applications in case an error condition arises and the two endpoints have to drop back to a known point. The third service is activity management, which allows the session layer to mark the beginning and end of activities and provide a quarantine service.

The remote procedure call allows a client application to request a service from a server by emulating the call of a remote procedure. Thus, similar to one local procedure calling a second local procedure with parameter passing, a client procedure calls a remote server procedure passing the appropriate parameters.

EXERCISES

1. Explain the difference between an end-to-end layer and a point-to-point layer.
2. List the three types of network as defined by the ISO.
3. List one example of each type of network as defined in question 2.
4. List the five classes of transport protocols.
5. Assuming an error-free network is desired, which transport protocol class would best provide this error-free network, given each of the three types of networks?
6. You are talking to a friend on the telephone and want to meet at a special place at 8:00. To make sure you both understand the correct location, give an example of a three-way handshake.
7. How many examples have we seen so far of flow control? List the protocol model and layer (if appropriate).
8. Explain the difference between upward multiplexing and downward multiplexing.
9. How does TCP recognize expedited data?
10. One of the problems of the remote procedure call was that a client could not use "pass-by-reference" parameters. What could the client use instead?
11. List as many scenarios as possible for what could happen if a client were to transmit a remote procedure call and the server were to crash before it could issue a response.

CHAPTER 10

PRESENTATION AND APPLICATION LAYERS

INTRODUCTION

The presentation and application layers are the final two layers of the OSI model. They are the "top" layers and as such are the farthest removed from the underlying hardware of a computer network. Unlike the session layer from the previous chapter, both the presentation and application layers can have significant effect on a network application. The presentation layer affects the presentation of data to the user of the application. This name, however, gives us an inaccurate picture. A number of professionals in the data communications field have stated quite wisely that this sixth layer of the OSI model should instead be named the representation layer, since the layer more accurately reflects the representation rather than the presentation of the data.

In this chapter we examine three common services provided at the presentation layer. The first service, **data representation**, introduces the Abstract Syntax Notation (ASN.1), used to transfer data structures from one network application to another. The second service, **data compression**, is employed to reduce the size of the data transferred in order to minimize the amount of data transmitted and the data transfer time. The third presentation service is **network security**. Specifically, we address data encryption and decryption and how it can be used to transfer data in a secure fashion.

The application layer creates an interface for the actual user/application programs that the network was designed to support. There are hundreds, perhaps thousands, of applications designed for computer networks. We obviously cannot list all of them. Luckily, several applications are widespread and found on many different types of networks. In the applications section we examine a number of the more common applications, including electronic mail, file transfer, and remote log-in.

DATA REPRESENTATION: THE ABSTRACT SYNTAX NOTATION (ASN.1)

In an early chapter we discussed how data is transmitted as a particular sequence of 1s and 0s depending on the choice of data code. How does one transmit something more elaborate than simple data, such as a data structure? If one network application wishes to transmit a data structure of information to another network application, what form does the data structure have during transmission? A common example of data structure transmission is the 24-hour bank card, or automatic teller machine. When you insert your 24-hour bank card and ask to withdraw $100 cash from the machine, a request is transmitted over a phone line to your banking institution. This request might include your name, account number, transaction type, amount, where the money is coming from (such as checking), and where the money is going (to the machine as cash). The **abstract syntax notation (ASN.1)** was designed to provide a protocol (literally a syntax) for transmission of possibly complex records; it consists of two parts: the description language, a Pascal-like code used to define the data structure; and the transfer syntax notation, used to transmit the data structure. ISO document 8824 defines the *abstract* syntax notation and ISO document 8825 defines the *transfer* syntax notation. While ASN.1 is so complicated that we could easily devote an entire chapter to it (an entire book would be better), we examine here only the salient features and explain them as simply as possible.

An ASN.1 structure is a collection of one or more primitives (or simple types). The most common primitives are these:

- BOOLEAN
- INTEGER
- REAL
- BIT STRING
- OCTET STRING—which can be further defined as NUMERIC STRING, PRINTABLE STRING, IA5STRING, GENERALIZED TIME, and others
- NULL—no type at all, but essentially a place holder
- ANY—a union of all possible data types
- OBJECT IDENTIFIER—while still a simple type, its description is fairly complex. Suffice it to say that an Object Identifier denotes an object that has been previously defined by an authoritative organization. The two organizations are CCITT and ISO (and a combination of the two organizations). Thus, if you wish to include in a particular data structure an object previously defined by either CCITT or ISO, you can use the Object Identifier to denote that object.

These simple types can be combined to create more complex types, called *constructed types*. There are five constructed types:

- SEQUENCE, an ordered list of various types, such as an Ada or Pascal record
- SEQUENCE OF, an ordered list of a single type, much like an array
- SET, an unordered collection of various types
- SET OF, an unordered collection of a single type
- CHOICE, the union of one or more data types

As an example, consider the following student record written in Ada:

```
type Student_Record is
  record
    ID : string(1..9);
    Name : string(1..20);
    Credit_Hours : integer;
    GPA : float;
    Currently_Enrolled : boolean;
  end record;
```

The corresponding ASN.1 syntax:

```
Student_Record ::= SEQUENCE {
  ID OCTET STRING,
  Name OCTET STRING,
  Credit_Hours INTEGER,
  GPA REAL,
  Currently_Enrolled BOOLEAN }
```

To provide flexibility to the actual transfer of data, any of the above types can be declared Optional. If a type is declared Optional, that data type can be omitted during transfer of the record. This raises an interesting question. If a data type can be optional, how does the receiver of the data type know which fields are included and which have been omitted? It might be obvious if we were transmitting a structure with widely different data types. But if we were transmitting a structure with 10 consecutive integers, we would not be able to tell which integer was missing if any were declared optional. We need some form of Tag that can be applied to each field to provide a form of identification. There are four types of tags: UNIVERSAL, APPLICATION, PRIVATE, and context specific. Universal tags are reserved for the primitive types defined by the ASN.1 standard (Boolean, Integer, Bit String, Octet String, and so on). Application tags can be employed to describe a unique data type within an abstract definition. For example, the above record Student_Record could be tagged with an Application tag. Private tags are unique within a given session, as agreed on by the two sides. Context specific tags are used to distinguish the various parts of a larger constructed type, such as identifying the fields of a Sequence of structure. Universal, Application, and Private tags have two parts: the type, followed by an integer. Only context-specific tags consist of just an integer.

As an example, let's place tags on our student record. A tag is denoted with brackets [ ].

```
Student_Record ::= [PRIVATE 0] IMPLICIT SEQUENCE {
  ID [0] IMPLICIT OCTET STRING,
  Name [1] IMPLICIT OCTET STRING,
  Credit_Hours [2] IMPLICIT INTEGER,
  GPA [3] IMPLICIT REAL,
  Currently_Enrolled [4] IMPLICIT BOOLEAN }
```

The record Student_Record has been tagged as a Private type, since its use has been agreed on by the users of this application. Each field within Student_Record has a context-specific tag, since its inclusion is related to the record Student_Record.

It is also possible to identify a data type as Implicit. If a data type is declared Implicit, then the tag is sufficient to identify the data type during transmission and it is not necessary to transmit its type also. Declaring types Implicit thus saves a few bytes of space during data transmission, as we shall see shortly when we convert to the transfer syntax notation.

Additional topics that could be discussed include defining Default values for simple types, allowing a type to be defined External to the current document, creating Subtypes of data types, assigning Values to the data types, and creating a Macro notation to further extend the power of ASN.1. If you are interested, you are encouraged to examine the Further Readings at the end of the book for more detailed information on these topic areas.

Having presented the description language for ASN.1, we can now examine how this description language is converted into the **transfer syntax notation**. Each value of the abstract syntax notation transmitted consists of three fields:

TYPE	LENGTH	VALUE(DATA)

The Type field consists of three subfields:

TAG	P/C	ID CODE

The Tag field is a 2-bit value and indicates the tag type for the ASN.1 item. A Universal tag is a binary 00, an Application tag is 01, a context specific tag is 10, and a Private tag is 11. The P/C bit is 0 if the type is a primitive type, and 1 if it is a constructed type. The Length field is the numeric length of the data field that follows. As an example, convert the student record example to its transfer syntax notation. First, however, we must assign some values to our student record:

```
Student_Record ::= [PRIVATE 0] IMPLICIT SEQUENCE {
    ID [0] IMPLICIT OCTET STRING,              "123456789"
    Name [1] IMPLICIT OCTET STRING,            "SMITH"
    Credit_Hours [2] IMPLICIT INTEGER,         42
    GPA [3] IMPLICIT REAL,                      3.45
    Currently_Enrolled [4] IMPLICIT BOOLEAN    TRUE}
```

The first statement to convert to the transfer syntax notation is

```
Student_Record ::= [PRIVATE 0] IMPLICIT SEQUENCE {
```

The Tag subfield for Student_Record = 11 (Private); it is a constructed type, so the P/C subfield = 1, and its ID Code = 0. This gives us a Type field of 11100000 (or hexadecimal E0). Since the data type Sequence is defined as Implicit, we do not need to create a tag for the data type Sequence (it would be a Universal tag). The Length of Student_Record is the length of all the following data types. Since we have not yet converted those data types, let's skip the Length field for now. The Value field for

Student_Record contains the values of all the following fields, which we will now convert.

The data type ID has been defined as context specific, so its Tag value = 10, its P/C value = 0 (primitive), and its ID CODE = 0. The resulting Type field is 10000000 (hexadecimal 80). The Length of ID = 9 (given the data "123456789"), and ID's Value equals the ASCII value of "123456789," or hexadecimal 31 32 33 34 35 36 37 38 39. The transfer syntax for data field ID is

```
TYPE    LENGTH    VALUE
80      09        31 32 33 34 35 36 37 38 39
```

Completing the remaining data types gives us the following values:

```
Name:                81  05  53  4D  49  54  48
Credit_Hours:        82  01  2A
GPA:                 83  02  7C  D4
Currently_Enrolled:  84  01  01
```

Name is the Type field (81) followed by the Length (05) followed by the ASCII characters for the name SMITH. Credit_Hours is the Type field (82) followed by the Length (01) followed by the hexadecimal value for 42. GPA is a real value and defies explanation at this time. Currently_Enrolled (True) has a Length of 01 and a Value of a nonzero value to denote True (a zero value would denote False). Now we can calculate the total Length of the data types ID, Name, Credit_Hours, GPA, and Currently_Enrolled = 28 (hexadecimal 1C), and insert this value as the Length field for Student_Record, giving the following complete transfer syntax notation:

```
E0 1C
        80  09  31  32  33  34  35  36  37  38  39
        81  05  53  4D  49  54  48
        82  01  2A
        83  02  7C  D4
        84  01  01
```

DATA COMPRESSION

Most of us are well aware of a basic tenet of calling someone on the telephone: the longer you talk, the more it is going to cost you financially. All long-distance phone calls are timed by your long-distance provider and you are charged a certain rate per minute. The local telephone call is still an untimed call in many areas, but it too is slowly becoming a timed call. Since the phone call charge is based on a per minute rate regardless of whether you are talking, it is not uncommon to hear the statement "We have to talk fast; this is a long-distance call" at the beginning of a telephone conversation.

As digital technology enters the telecommunications market, we are beginning to see a change in how line usage is computed. Instead of charging by a unit of time, charges are computed by the number of bits or bytes transmitted. This is possible since digital transmission uses discrete pulses of voltage that can easily be counted. Thus, it does not matter how fast one transmits data over a line that counts the number of bits;

the line usage charge would be the same. The answer to a more economical use of the transmission line is to transmit fewer bits. To convey the same meaning of the data while transmitting fewer bits, it is necessary somehow to compress the data. There are two basic techniques for compression of transmitted data: **frequency-dependent compression** and **context-dependent compression**.

Frequency-Dependent Compression

In Chapter 2 we examined the EBCDIC and ASCII data codes that are used to represent the symbols of a character set. While some codes use a 7-bit representation, and other codes use an 8-bit scheme, all the codes examined thus far have one characteristic in common: all symbols transmitted using a particular code have the same bit length. If we are transmitting 100 characters using an 8-bit code, we will transmit 800 bits (not counting extraneous bits for framing, error checking, and so on). But this does not always have to be the case. If there were a way to identify the more frequently transmitted characters, and we represented those characters with shorter bit codes, we should be able to transmit fewer bits than when using a fixed method. This is the concept behind **Huffman Codes**. In a Huffman Code, the more frequently occurring characters are represented with shorter bit patterns while the less frequently occurring codes are represented with longer bit patterns.

As an example, consider the following situation. Suppose we have a small data set with only eight possible characters, A through H. We could choose a fixed method and use a 3-bit code to represent each of the eight characters. Character A would be 000, B would be 001, C 010, and so on. If we transmitted 100 characters using this scheme, we would send 300 bits. Now, suppose the eight characters occur with the following frequencies: A:6%; B:3%; C:18%; D:32%; E:2%; F:24%; G:5%; and H:10%. To create a Huffman Code for these eight characters, list the characters in increasing order of frequency as in Code 1:

E	B	G	A	H	C	F	D	
2	3	5	6	10	18	24	32	(frequencies)

Code 1

Take the two characters with the smallest frequencies and combine them, adding their frequencies together, as in Code 2:

Code 2

Continue combining two characters at a time, always selecting the two characters with the smallest frequencies. If two previously combined characters have a combined value less than a single character, combine that combination with the other smallest frequency, as shown in Code 3 at the top of the next page. At this point we would like to combine 26 and 32, the two smallest frequencies, but that would cause us to cross lines when we next try to combine 42. To avoid the crossing, we rearrange the characters so that D is between H and C, and continue combining, as shown in Code 4, immediately following Code 3.

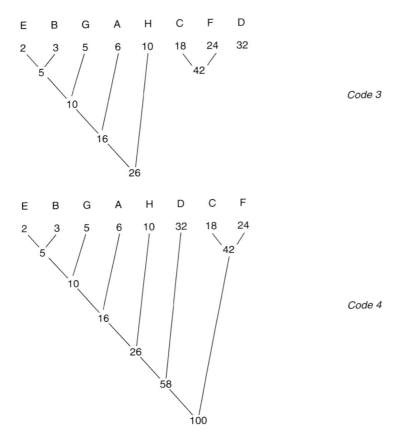

Code 3

Code 4

Now that all frequencies are combined with other frequencies, assign a 0 bit to a left arc and a 1 bit to a right arc, starting at the bottom of the graph (see Code 5 below).

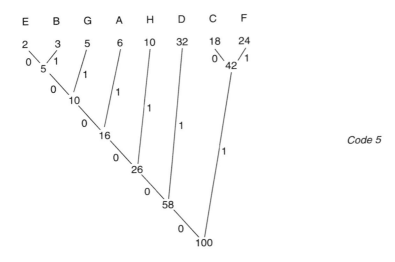

Code 5

The following character codes will result: A = 0001; B = 000001; C = 10; D = 01; E = 000000; F = 11; G = 00001; and H = 001.

To estimate the number of bits transmitted using the previous character codes, multiply the number of bits in each character by its original percentage of occurrence:

A = 4×0.06; B = 6×0.03; C = 2×0.18; D = 2×0.32; E = 6×0.02;

F = 2×0.24; G = 5×0.05; and H = 3×0.10.

Adding all values yields 2.57 bits per character. Thus, if 100 characters were transmitted using this Huffman Code, only 257 bits would be sent, saving 43 bits over the earlier fixed method.

Context-Dependent Compresssion

The second technique for performing data compression relies on the conditional probability of a particular character following another character. Consider, for example, the digitization of a black-and-white picture, such as a photograph in a newspaper. If you look closely at the photograph, you see that it is composed of tiny black-and-white dots. If we let a binary 1 represent a black dot and a binary 0 represent a white dot, we could represent the black-and-white photo as a long series of 1s and 0s. To store or transmit the photo, we would transmit this long string of 1s and 0s. Often, however, the background of a photograph is all one color, such as black, making the binary 1 the predominant bit in the data stream. We could use this knowledge to compress the bit stream, reducing the quantity of 1s.

Given the following bit string, we see relatively long successions, or "runs," of 1s:

11111011110111111111101111110111001111111111111111101111111

The first step to compressing this data is to count the number of 1s in each run. In this example, there are five 1s, a 0, four 1s, a 0, ten 1s, a 0, six 1s, a 0, three 1s, a 0, zero 1s, a 0, sixteen 1s, a 0, and seven 1s. Considering only the runs of 1s, we have this set:

5 4 10 6 3 0 16 7

We next convert each value into its 3-bit binary equivalent value, with any value larger than 7 represented as successive groups of 3-bit values:

5 = 101; 4 = 100; 10 = 111 011; 6 = 110; 3 = 011; 0 = 000;

16 = 111 111 010; and 7 = 111 000.

The final transmitted bit string is all the 3-bit binary values:

101   100   111   011   110   011   000   111   111   010   111   000.

We have compressed 58 bits down to 36 bits for a 38 percent reduction. It is not uncommon to find a context-dependent compression scheme that can reduce the original data stream by as much as 50 percent.

Audio and Video "Compression"

Throughout this text, as well as other texts on data communications, much emphasis is placed on accurately transmitting data from one point to another. Receiving data that is "close" in content to the originally transmitted data is typically not acceptable. If even a single bit is altered during transmission, the resulting data stream could be invalid, forcing a retransmission of the data. Compressing data into a smaller package requires the same amount of accuracy. The compression technique employed must decompress the compacted data, returning it to its original form, without the loss of a single bit. You would not want to perform a bank transaction such as transferring money from one account to another if the result is only "close" to what you had authorized.

As we continue into the digital age of computers, we are also experiencing the digitization of other forms of data that have traditionally been analog. Music and video are two good examples of analog formats that are now stored, transmitted, and reproduced in digital form. The compact disc, followed by the digital compact cassette, digital audiotape, and minidisk formats, all store analog music using a series of binary 1s and 0s. While the bulk of video is still stored and transmitted using analog means, more examples of digital video appear every day.

The major problem with storing music and video in digital form is the large amount of data that is produced. Compact discs can store approximately 75 minutes of music, which is more than 600 megabytes of digital storage. The digital compact cassette and the minidisk formats cannot store 600 megabytes of data, thus they cannot hold the contents of one music album. For them to do so, some form of data compression is needed.

Digital video is even more problematic. One frame of a video picture when digitized requires almost 1 megabyte of storage. Consider that when viewing full-motion video you are seeing 30 frames per second times 60 seconds per minute times 120 minutes for a typical movie, and you have over 215,000 *megabytes* of storage, or roughly 360 compact discs. Some form of compression is clearly needed if we are going to store a two-hour video in digital form.

Luckily, there are two big differences between audio/video data and computer data. First, audio and video data start in analog form. If audio or video data is to be transmitted or stored in a digital format, **analog-to-digital conversion** must be performed on the original data. Some amount of accuracy is almost always lost when analog data is converted to digital. Second, audio and video data do not have to be reproduced exactly as they originally appeared. They only have to be close. This is because the average human ear and eye perceive differing examples of music and video as similar. It is possible to compress music, removing certain sounds that the average listener does not typically hear anyway. With this **perceptual coding** it is possible to compress the contents of a record album to fit a digital compact cassette or a minidisk. Video compression is similarly clever, and some very advanced techniques are used to compress a movie (the equivalent of 360 compact discs' worth of digital video) down to just two compact discs!

Data compression, however, results in data being reduced to a more compact size but still retaining the exact original content. Since this is not the intent of audio and video "compression," one should instead call it **audio and video reduction,** to distinguish it from data compression.

ENCRYPTION AND DECRYPTION_____

Many times when data is transferred from one point to another, it is necessary to ensure that the transmission is secure from anyone eavesdropping on the line. The term *secure* implies two conditions: it should not be possible for someone to intercept and copy an existing transmission; it should not be possible for someone to insert false information into an existing transmission. Banking transactions and military transmissions are two examples of data that should be secure during transit.

The introduction of fiber optic cable was a major improvement in providing secure data transmission. Metal-conducted media that are transmitting electrical signals are relatively easy to tap into, but fiber optic media using light pulses are much more difficult to tap. Of course, no system by itself is secure from an individual determined to steal data, thus additional measures are necessary. One such measure is to use software to encrypt the data before transmission, transmit the data over secure media, then decrypt the received data to obtain the original information. Extensive research has been conducted on **cryptography** by the government, universities, and private industry. Many volumes can be filled on cryptographic techniques, but we examine only the simpler techniques here.

Before we jump into the exciting topic of cryptography, a few terms should be introduced. **Plaintext** is the name applied to the data before any encryption has been performed (Figure 10.1). The plaintext is given to an **encryption algorithm** that uses a **key** to create the **ciphertext**. The ciphertext is transmitted to the receiver, where the ciphertext data is given to a **decryption algorithm**. This decryption algorithm uses a second key to decrypt the ciphertext, producing the original plaintext. Early cryptography techniques required the two keys to be the same. Newer techniques allow the use of two different but mathematically related keys. For our example, plaintext is shown in lower case while ciphertext is shown in uppercase.

Monoalphabetic Substitution-Based Ciphers

Despite the daunting name, monoalphabetic substitution-based ciphers represent a fairly simple ciphering technique. The idea is to replace a character or group of characters with a different character or group of characters. Consider the following. Each letter in the plaintext row maps onto the letter below it in the ciphertext row.

Plaintext: a b c d e f g h i j k l m n o p q r s t u v w x y z

Ciphertext: P O I U Y T R E W Q L K J H G F D S A M N B V C X Z

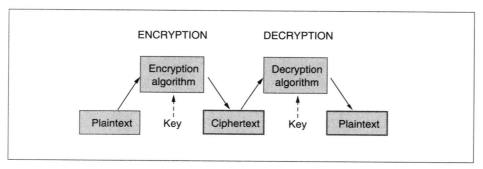

FIGURE 10.1   Basic encryption/decryption procedure

The ciphertext chosen is simply the letters on a keyboard, scanning right to left, top to bottom. To send a message using this encoding scheme, each plaintext letter of the message is replaced with the ciphertext character directly below it. The message "how about lunch at noon" would encode to "EGVPO GNMKN HIEPM HGGH." A space has been placed after every five ciphertext characters to help disguise obvious patterns. This example is monoalphabetic since one alphabetic string is used to encode the plaintext, and is a substitution cipher because one character of plaintext is substituted with one character of ciphertext.

Polyalphabetic Substitution-Based Ciphers

The polyalphabetic substitution-based cipher is similar to the monoalphabetic cipher except for one interesting wrinkle: rather than use one alphabetic string to encode the plaintext, multiple alphabetic strings are used. Possibly the earliest example of a polyalphabetic cipher is the **Vigenére Cipher** created by Blaise de Vigenére in 1586. A 26 x 26 matrix of characters is created where the first row, the A row, equals ABCDEFGHIJKLMNOPQRSTUVWXYZ, the second row, the B row, equals BCDEFGHIJKLMNOPQRSTUVWXYZA, the third row, the C row equals CDEFGHIJKLMNOPQRSTUVWXYZAB, and so on. A key is chosen, such as COMPUTER SCIENCE, which is repeatedly placed over the plaintext message. For example:

Plaintext: C O M P U T E R S C I E N C E C O M P U T E R S C I E N C E C O M P U T E R S C I E N C E C O

Ciphertext: t h i s c l a s s o n d a t a c o m m u n i c a t i o n s i s t h e b e s t c l a s s e v e r

To encode the message, we take the first letter of the plaintext, t, and the corresponding key character immediately above it, C. The C tells us to use row C of our 26 x 26 matrix to perform the alphabetic substitution of the plaintext character t. The ciphertext character V would be encoded in t's place. This process continues for the every character of the plaintext. The key, of course, should be kept secret between the encoder and decoder, which always introduces additional problems.

To make matters more difficult for an intruder, the standard 26 x 26 matrix with A row, B row, C row, and so on does not have to be used. However, if a unique matrix is chosen for the encoding and decoding, it too must remain a secret along with the key, which makes things a little more difficult for the encoder and decoder.

Transposition-Based Ciphers

A transposition-based cipher is different from a substitution-based cipher in that the order of the plaintext is not preserved as it is in substitution ciphers. By rearranging the order of the plaintext characters, common patterns become unclear and the code is much more difficult to crack. To create a transposition cipher, choose a keyword that contains no duplicate letters, such as COMPUTER. Over each letter in the keyword, write the position of each letter as it appears in the alphabet. For the keyword COMPUTER, C appears first in the alphabet, E is second, M is third, O is fourth, and so on.

```
1   4   3   5   8   7   2   6

C   O   M   P   U   T   E   R
```

Now, take the plaintext message (this class is the best class i have ever taken) and write it under the keyword in consecutive rows going from left to right.

1	4	3	5	8	7	2	6
C	O	M	P	U	T	E	R
t	h	i	s	c	l	a	s
s	i	s	t	h	e	b	e
s	t	c	l	a	s	s	i
h	a	v	e	e	v	e	r
t	a	k	e	n	i	n	m
y	l	i	f	e			

To encode the message, read down each column starting with column 1 proceeding through column 8. Reading column 1 gives us TSSHTY and column 2 is ABSEN. Encoding all eight columns gives us the following message:

> TSSHTYABSENISCVKIHITAALSTLEEFSEIRMLESVICHAENE

Once again, the choice of the keyword is important and care must be taken that it does not fall into the wrong hands.

Key Problems

All the encoding and decoding techniques shown thus far depend on protecting the key and keeping it from an intruder. Yet, as important as key secrecy is, it is surprising how often keyword or password security is lax or nonexistent. Consider the episode of Stanley Mark Rifkin and the Security National Bank. Posing as a bank employee, Rifkin gained access to the wire funds transfer room, found the password taped to the wall above the computer terminal, and transferred $12 million to his personal account.

One technique to protect a key from an intruder is often seen in a late-night black-and-white spy movie: break the key into multiple pieces and assign each piece to a different individual. Rather than simply assign one or two characters to each person, use mathematical techniques such as simultaneous linear equations to divide the key into parts.

One of the problems with protecting a key is that only one key is used both to encode and decode the message. The greater the number of people who have possession of the key, the greater is the possibility that someone will get sloppy and disclose the key to unauthorized personnel. If two keys are used, one to encode the plaintext and a second to decode the ciphertext, it is easier to protect the message from intrusion. It would be even more powerful if the encryption key could be public knowledge, thus eliminating the need for encryption key security. From the work of Diffie and Hellman comes the concept of **public key cryptography**. Two algorithms are used with public key cryptography, an encryption algorithm E and a decryption algorithm D. Even though the E algorithm encrypts the plaintext and the D algorithm decrypts the corresponding ciphertext, if one has access to the E algorithm, it is extremely difficult to deduce the D algorithm. Furthermore, the E algorithm is also extremely difficult to break using a plaintext attack.

APPLICATION LAYER

Electronic Mail

One network application that has received widespread use is electronic mail. Many different examples of electronic mail can be found over the entire spectrum of computer networks. At one installation, it is not uncommon to find one electronic mail system that operates on a local area network, a second system that operates on the site mainframe computer, and then access to a third electronic mail system such as the Internet SMTP mail system or the educational BITNET mail system.

While each electronic mail system is unique, most have some basic operations in common:

- Ability to compose, edit, and send a mail message
- Ability to receive and file a mail message
- Ability to forward a mail message to a third party
- Ability to send carbon copies and blind carbon copies
- Ability to assign priorities to a mail message
- Ability to assign an expiration time to a mail message
- Ability to send a message to multiple parties by a distribution list
- Ability to inform sender of message arrival confirmation

In an attempt to standardize electronic mail, CCITT established a series of X.400 Message Handling Systems (MHS) standards in 1984. In 1988 ISO came out with version, MOTIS, which was based fairly closely on the X.400 series. Briefly, MHS defines a collection of services between the two end-users of an electronic mail system: the originator and the recipient. The services are broken into two categories: the message transfer system (MTS), which is responsible for the "envelope" or the details surrounding the message; and the interpersonal messaging (IPM) services, which are responsible for the contents of the "envelope," or the message itself. The Message Handling Systems recommendation describes the components of the electronic mail system as **functional objects**. The functional objects, such as the **User Agent**, the **Message Transfer Agent**, the **Message Store**, and the **Access Unit**, interact with one another passing envelopes through the system (Figure 10.2).

The User Agent allows a user access to the electronic mail system. It is responsible for preparing, sending, and receiving messages as well as providing levels of priority, security, and delivery notification. The Message Transfer Agent performs the relaying or routing of the envelopes. The Message Store, as its name implies, provides a storage area for envelopes, or messages. The Access Unit provides support for connecting the electronic mail system to any outside services, such as other mail systems or postal services. The X.400 recommendations also provide the definition of approximately 90 MHS **Service Element**s. These service elements form the core of MHS by defining the

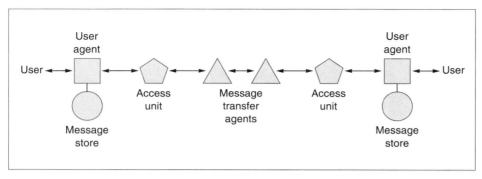

FIGURE 10.2 Functional objects of an electronic mail system

features and functions of a particular mail system. Examples of service elements include auto-forward indication, express mail service, message identification, proof of delivery, and special delivery, to name only a few.

File Transfer

Most networks provide some means of transferring a file from one computer to another. While this may sound like a simple procedure, it is not. The concept of a general purpose file transfer protocol includes so many details that attempts at defining a standard have been very difficult. The file transfer procedures that exist today are usually specialized to a particular network, or brand of network. Three examples of file transfer protocols that come close to working on a wide variety of networks are the Internet's FTP, OSI's FTAM, and the public domain program KERMIT. Let's spend a few minutes examining each of these.

FTP (File Transfer Protocol) is a file transfer and access program that originated with the older ARPANET network and operates on top of the TCP/IP suite of protocols. Since FTP is based on TCP/IP, it is associated with the Internet and is actually its most used application. A user wishing to transfer a file using FTP engages in an interactive dialogue that allows for user authentication, file format control, remote directory inquiry, and file transfer commands. The actual format for file data transfer across the network is based on another well-known standard, TELNET. TELNET is a TCP/IP network virtual terminal protocol and is described in more detail in the following section.

The basic FTP model creates a client control connection when a client, or user, enters an FTP session on its host computer. As long as the user is in an FTP session, the client control connection exists. During this period the user logs into the remote computer to/from which the file is to be transferred. The user may also specify options such as binary image file transfer or ASCII character file transfer. When the user issues the get/put command to receive/send a file, a data transfer connection and data transfer process are created at both the client and server locations. When the file has completed the transfer across the network, the data transfer connection and process are deleted.

As an example, consider the following FTP dialogue between a user and the host computer operating system:

$FTP—the user issues the local FTP command and enters an FTP session

ftp>open id.loc.net—user opens a connection with id.loc.net host

ftp>Name:—user enters account name (or ANONYMOUS)

ftp>Password:—user enters appropriate password

ftp>dir—user lists directory

ftp>cd pub—user changes to public directory

ftp>get filename—user requests "filename" be copied to local account

ftp>close—user closes connections

ftp>exit—end FTP session with local computer

FTAM (File Transfer, Access, and Management) is the attempt by ISO to create a generic file transfer standard for an OSI-type network. While the details of FTAM are numerous and are contained in the ISO 8571 document, we discuss here a few of the more interesting features. FTAM is a connection-oriented process, and a user, when creating a session with FTAM, enters the FTAM Association Regime. Once the regime is entered, the user may execute one of many "primitives," such as creating a file, selecting a file for transfer, reading the file attributes, and transferring the file.

FTAM also allows for creating reliable file transfers, recovery from network failures, resynchronization after minor network perturbances, the establishment of periodic checkpoints during file transfer, and the ability to bracket multiple primitives together into one atomic operation. While many of these features sound reasonable, ISO attempted to create a file transfer protocol that was as generic as possible, and in the process, created a procedure that is very large and difficult to use, at best.

KERMIT (which surprisingly is not an acronym) is a file-transfer program that was designed to transfer files between microcomputers and mainframes. It is the creation of Frank da Cruz and Bill Catchings of Columbia University. Their goal was to create a file transfer mechanism that transfers ASCII, EBCDIC, or binary files over common asynchronous serial transmission lines. Furthermore, the transfer of files occurs over standard terminal connections in half-duplex stop-and-wait acknowledgment form, and the data packets are limited to a maximum length of 96 characters. Only the file's name and contents are transmitted, not the file attributes.

The contents of each data packet begin with a control-A (BISYNC Start of Header) character, followed by a one-character LENgth field that denotes the length of the remaining packet. Following the LENgth field is the SEQuence field, with values from 0 to 63, which indicates the sequence of this packet in the file transfer stream. The next field is the Type field, which indicates what type of packet this is, such as D(ata), Y(acknowledge), N(egative acknowledge), and others. Following the Type field comes the Data portion of the packet—a stream of ASCII characters. Nonprintable ASCII characters are prefixed with special characters and then converted to printable characters by complementing the seventh bit. The final field is the CHECKsum field, which can be a one-, two-, or three-character arithmetic or cyclic checksum.

As a simple example, consider the following sequence of packets. Note the similarity between this dialogue and a dialogue from a BISYNC conversation. The exact contents of each packet are not shown for brevity:

```
Send Initiate Packet →
← ACK for Send Initiate
```

```
File header containing name of file →
← ACK for file header
First data packet →
← ACK for first data packet
Second data packet →
← ACK for second data packet
Third data packet →
← NAK for third data packet; it was received garbled
Third data packet →
<- ACK for third data packet
    :
    :
End-of-File packet sent →
← ACK for End-of-File
End-of-Transmission packet sent →
← ACK for End-of-Transmission
```

Remote Log-in

The final common network application examined here is the ability to login to a host computer from a remote site. One of the more popular programs that allows this feature is **TELNET**. TELNET is a relatively simple remote terminal protocol that operates over the TCP/IP suite of protocols. TELNET allows a user at a remote site (client) to establish a TCP connection to a log-in server at a second site, and then transmit terminal keystrokes to the log-in server as if the remote site were a terminal connected directly to the second site.

Three basic services are offered by TELNET:

- TELNET defines a network **virtual terminal** service that can provide a standard interface to remote systems.
- TELNET allows client and server to negotiate options, and provides a set of standard options.
- Both ends of the TELNET connection are treated equally, allowing both the client and the server to negotiate options.

Since there is such a variety of terminals in existence, and different terminals may use different character sequences to perform terminal functions, it is difficult to interface heterogeneous types of terminals. To allow different types of client terminals to communicate with the TELNET server, all transmitted keystrokes are converted to a common form using the **network virtual terminal** (**NVT**) format. The NVT format uses the standard seven-bit ASCII character set for data transmission and requires various character-code sequences such as CR-LF (carriage return–line feed) for the end-of-line transmission sequence.

Option negotiation is possible, allowing client and server to tailor the transmission line. Negotiable options include changing transmission to 8-bit binary, exchanging information about the make and model of a terminal being used, allowing one end to echo received data, and requesting the status of a TELNET option from a remote site.

To negotiate an option, a client or server issues a Will X command, meaning, "Will you agree to let me use option X?" The receiving side can then issue a Do X or a Don't X command in response.

CASE STUDY

Back at XYZ University, the computing center staff has finally installed most of their workstations, modems, cables, and support peripherals for the campus computer network. Numerous departments on campus are already using the local area networks and backbone network for completing homework assignments, classroom demonstrations, administrative activities, and conversing with the mainframe computer. If the computing center staff had any fears the campus network would be underutilized, those fears are quickly vanishing.

While the computing center staff do not see a present need to create new applications for individual university departments, they want to provide a host of services that all departments on the campus may use. For example, because of the large volume of transmitted data, a number of applications could benefit from the presentation service of data compression. Any application that deals with the transmission of graphic images would indeed realize a substantial transmission improvement if the graphic image were compressed before being transmitted. Several software programs exist which can apply one or more compression techniques to lengthy data files. If data is to be transmitted through modems, like the connection between XYZ University and State University, modems can be installed that also provide data compression.

Various applications on campus may also contain confidential data. To ensure that the data remains secure and is not intercepted by unauthorized personnel, the computing center can apply data encryption/decryption software before the sensitive data is transmitted over a network or phone line. The computing center staff can select from an abundance of security software packages.

Since many departments will want to establish an electronic mail service for use by their employees, it seems worthwhile for the computing center to encourage installation of a campuswide electronic mail system. The staff will undoubtedly follow a standard systems analysis of investigation, analysis, design, implementation, and operation and maintenance. The potential users of a proposed E-mail system should be consulted for their input, which should be converted to a list of product requirements. Several E-mail software products will be examined, the best will be selected, installed, and tested, and all potential users of the system will be offered training. Documentation that clearly explains details and demonstrates the essential functions of the software will be provided. Finally, the computing center staff will commit to the continued maintenance of the chosen system.

The computing center staff will also want to install the appropriate file transfer software and remote log-in software necessary to allow approved campus users access to the services of the Internet. Training will be necessary, as well as clearly written documentation.

As always, the staff members of the computing center will continue to stay abreast of the evolving technology by reading current industry literature, taking classes, and attending conferences, symposiums, and trade shows.

SUMMARY

The presentation and application layers are the final two layers of the OSI model. The presentation layer affects the representation of data to the user of the application. The application layer creates an interface for the actual user/application programs the network was designed to support.

The abstract syntax notation (ASN.1) was designed to provide a protocol for transmitting possibly complex records; it consists of two parts: the description language, a Pascal-like code used to define the data structure; and the transfer syntax notation, used to transmit the data structure.

A second presentation service commonly provided is data compression. Two basic techniques for compression of transmitted data were presented: frequency-dependent compression and context-dependent compression. When using a frequency-dependent compression technique such as a Huffman Code, the more frequently occurring characters are represented with shorter bit patterns, while the less frequently occurring codes are represented using longer bit patterns. Context-dependent compression relies on the conditional probability of a particular character following another character, such as the digitization of a black-and-white picture.

Encryption and decryption of data allow us to create a secure data transmission line. The term *secure* implies two conditions: an intruder should not be able to intercept and copy an existing transmission; an intruder should not be able to insert false information into an existing transmission. Among the encryption devices used to thwart intruders are three kinds of ciphers. A monoalphabetic substitution-based cipher replaces a character or group of characters with a different character or group of characters. The polyalphabetic substitution-based cipher is similar to the monoalphabetic cipher except that multiple alphabetic strings are used to encode the data. A transposition-based cipher rearranges the order of the plaintext characters so that common patterns become unclear and the code is much more difficult to crack.

One network application that has received widespread use is electronic mail. There are many kinds of electronic mail operating over the entire spectrum of computer networks. A second common network application is file transfer. Most networks provide some means of transferring a file from one computer to another. The third common network application examined is the ability to log-in to a host computer from a remote site.

EXERCISES

1. ASN.1 has primitive types and constructed types. List three examples of each.
2. What problems arise when data types are declared Optional in the ASN.1 syntax?
3. Convert the following Ada record into equivalent Abstract Syntax Notation and Transfer Syntax Notation.

```
type Customer_Rec is
  record
    LName : string (1..10);
    FName : string (1..10);
```

```
    Credit  : float;
    AcctNo  : string (1..7);
    PastDue : boolean;
    Address : string (1..20);
  end record;
```

4. Create a Huffman Code for the symbols 0 through 9, using the following percentages of symbol occurrence: 12, 7, 18, 4, 19, 8, 3, 20, 1, and 8, respectively.
5. How many bits were saved by transmitting 100 characters using the Huffman Code created in question 4?
6. Using a context-dependent compression scheme, compress the following bit stream:
   11111110111110111111010111100111110111111111111111101111110111
7. What is the major difference between music/video compression techniques and computer data compression techniques?
8. Create a monoalphabetic substitution-based cipher and encode the message "meet me in front of the zoo." Explain your encoding technique.
9. List five operations common to all basic electronic mail systems.
10. How is the Kermit file transfer technique similar to the BISYNC protocol? Validate your answer with an example.

EMERGING TECHNOLOGIES

INTRODUCTION

Trying to predict the future of an industry is much like gambling at Las Vegas: sometimes you win, but mostly you lose. Predicting the computer industry, depending on whom you talk to, is either easier or harder. For instance, it is a fairly safe bet that electronic components will continue to shrink in size, increase in speed, and drop in price. It is also fairly safe to assume that the demand for computers and computer-related equipment will generally continue to increase. Surely the demand for quality software will continue, as good software always lags behind hardware in development.

The difficult part of predicting the future of the computer industry is trying to predict what new products will be developed, whether they will be accepted by consumers, and how existing products may someday be used in new applications.

In this final chapter, we don't try to predict the future of the data communications industry. Instead, we examine a number of technologies that have been introduced within the past few years. Normally, technology that is two to three years old is often *old* technology and is soon to be replaced, if it hasn't been replaced already. One unique characteristic of a number of the technologies introduced in this chapter is their immensity. Rather than a team of three or four people creating a new device or software in a couple months, hundreds, perhaps thousands, of scientists, engineers, technicians, programmers, industry leaders, government agencies, and users have labored for years to bring these new technologies to market.

A second characteristic of the technologies introduced here is the level of difficulty they present to someone trying to understand their technical details. Essentially every topic we discuss could have a chapter or two devoted to it, and in some cases, an entire book. We begin with one of the largest and most ambitious projects currently in existence: the **Integrated Services Digital Network** (**ISDN**).

INTEGRATED SERVICES DIGITAL NETWORK (ISDN)

ISDN was designed in the mid-1980s to provide a worldwide public telecommunication network that would support telephone signals, data, and a multitude of other services for both residential and business users. A user would be connected to ISDN via a "digital pipe." This pipe, which consists of only one or two pairs of twisted pair wire, could provide the user access to circuit switched networks, such as telephone networks, packet switched networks, such as public data networks, and a host of local area network services. All these services could be provided simultaneously, at varying data transfer rates, and the user would be charged according to the capacity used, not the connect time.

Furthermore, ISDN was designed with a number of specific objectives in mind:

■ ISDN would provide access to both dial-up and leased telephone services.
■ There would be one single set of ISDN standards.
■ The network could provide a separation of competitive services.
■ One type of service would not subsidize another service.
■ There would be a smooth migration from the current telecommunication system to ISDN.

The types of services provided by ISDN would include the following:

■ Facsimile
■ Videotex
■ Teletex
■ Alarm systems
■ Many modern telephone services, including call transfer, caller identification, caller restriction, call forwarding, call waiting, hold, conference call, and credit-card calling

The digital pipe that connects the user to the ISDN central office would carry a combination of various types of channels. Three basic kinds of channels have been defined thus far:

■ **B channel**—basic user channel—is a 64 Kbps channel capable of carrying digital data, PCM-encoded voice, or a mixture of lower-rate traffic. Three types of connections can be established using a B channel: circuit switched connections (dial-up telephone); packet switched connections (PDN X.25 service); and semi-permanent connections, or leased lines.
■ **D channel**—data traffic channel—is a 16 or 64 Kbps channel that would carry the signaling information for the B channel. The D channel may also be capable of carrying packet switched data or low-speed data such as emergency services information.
■ **H channel**—high-speed data channel—is a 384 Kbps channel (H0), 1.536 Mbps channel (H11), or 1.92 Mbps channel (H12) that is capable of carrying high-

speed data for applications such as facsimile, video, high-quality audio, or multiple lower-speed data streams.

A user wishing to subscribe to ISDN would receive a "package" that would consist of combinations of the above three channels. Different combinations could be created by the ISDN provider for varying user requirements. For example, a "basic service" package could be created consisting of two full-duplex B channels plus one full-duplex 16 Kbps D channel. A basic service package would be standard for most residential services and many small business services. It would allow the simultaneous transfer of voice as well as a number of data applications.

For users requiring a larger data capacity, the "primary service" package could be used. The primary service package could consist of the equivalent of 23 B channels plus one 64 Kbps D channel for a combined data transfer rate of 1.544 Mbps (T-1 speed). Additional services have also been defined which include the H channels in various configurations with the B and D channels.

While the average user has probably never heard of ISDN, many telecommunication and business companies in the United States are migrating toward this standard. However, because of the extremely large number of standards involved, the migration will not be a quick one and will require a number of years before widespread consumer use is noticed. In Europe, however, ISDN has a very strong base, and it continues to be a major player in wide area network development in many countries.

BROADBAND ISDN

Shortly after the introduction of ISDN, engineers and others realized that various applications would quickly surpass its available bandwidth. Thus, in 1988 CCITT introduced the first of its recommendations for **broadband ISDN**, or **BISDN**. One of the primary differences between the older "narrowband" ISDN and broadband ISDN is the data transfer rates: broadband ISDN is defined to support data rates over 600 Mbps. Since narrowband ISDN twisted pair is not capable of supporting such a high data rate, broadband ISDN was designed for fiber optic cable. A new switching technology had to be introduced to accommodate the high data transfer rates of broadband ISDN: asynchronous transfer mode (ATM), which is discussed in the following section.

Three new transmission services have been defined for broadband ISDN:

■ A full-duplex 155.52 Mbps service
■ An asymmetrical service that provides a 155.52 Mbps transmission stream from the user to the network, and a 622.08 Mbps transmission stream from the network to the user
■ A full-duplex 622.08 Mbps service

The first service of 155.52 Mbps should easily support any requirements of an existing narrowband service. The second service with the higher 622.08 Mbps inbound rate and the slower 155.52 outbound rate will work well with systems that provide a video display to a user who does not need to respond with an equivalent video transfer display. The third service will handle multiple video sources, inbound and outbound, such as videoconferencing.

All broadband ISDN traffic will consist of packets of information, similar to the X.25 protocol of public data networks. Unlike X.25, however, all broadband ISDN packets will be the same size, and any signaling information will be transmitted on a separate common signaling channel (like narrowband ISDN's D channel).

ASYNCHRONOUS TRANSFER MODE (ATM)

The **Asynchronous Transfer Mode (ATM)** was chosen for the transmission technique in broadband ISDN because of its simplicity, its ability to transfer data rapidly between two points, and its ability to handle varying rates of data quickly. All data is packaged into fixed-sized units, called *cells*. Each cell consists of a 5-byte header followed by a 48-byte data field. Similar to X.25, ATM allows the multiplexing of numerous logical connections over a single physical connection. Unlike X.25, however, there is no link-by-link error control nor is there any flow control information provided within a cell. The developers assumed that since ATM will operate over reliable, modern fiber optic networks, a minimum of error and flow control will be necessary at the cell level. Any desired error control could be included at a higher layer in the communications software.

FIBER OPTICS

The use of fiber optic cables for all modes of communication transfer continues to expand every day. Many long-distance telephone lines have been replaced by either fiber optic cable or microwave transmission. Local area networks now use fiber optic interconnections as well as a number of high-speed mainframe computer interconnections. When fiber optic cables first began replacing metal cables, the applications continued to use protocols and standards that did not take full advantage of the superior characteristics of fiber optic. Eventually, standards such as FDDI appeared, designed specifically for fiber optic capabilities.

Two of the most recently created standards that employ fiber optic cables are **Synchronous Optical NETwork (SONET)** and **Synchronous Digital Hierarchy (SDH)**. SONET was proposed by BellCore and standardized by ANSI, while SNH is a CCITT standard. Both formats define a synchronous multiplexing format for transmitting low-level digital signals over fiber optic cable. Both SONET and SDH are fairly advanced and have very involved standards. Their explanation is beyond the scope of this text, but the key points to understand are these:

- Fiber optic cables are here to stay, not only in data communication transmission and telephone systems but in home entertainment systems, government and business applications, and even the automobile
- New standards, which fully exploit the capacities of fiber optic cable, will continue to be created.

WIRELESS TECHNOLOGY

The growth of wireless technology can best be described as explosive. Some industry specialists predict that wireless technology will have the same impact on society as did the telephone, radio, and television. Wireless systems can also be categorized as **Personal Communications Services (PCS)** and **Personal Communications Networks (PCN)**. PCS and PCN include many of the following innovations:

- *Cordless telephones*—Originally analog, but quickly becoming digital, cordless telephones include not only the home cordless sets that operate only within a couple hundred feet of the base, but also the more powerful units that can operate over hundreds of meters within large buildings. A number of companies now market a telephone, which, when it is within reach of the home base unit, acts like a cordless telephone. However, when the telephone is outside the reach of the home base unit, it automatically becomes a mobile cellular telephone and communicates with the local cellular phone system.

- *Mobile cellular radio (telephone)*—Originally analog and also quickly becoming digital, mobile cellular radio used to exist only in specialized markets for unique applications. Now cellular radio networks extend across much of the United States and include numerous forms of mobile units, including automobiles, emergency vehicles, and pedestrians.

  As the cellular radio units become smaller and more lightweight, they are being combined with other forms of electronic devices. For example, the Bell-South telephone company markets a handheld personal digital assistant that incorporates a cellular radio. Thus, it not only provides computer functions such as an electronic memo pad, datebook, calculator, and filing system but can also send facsimiles, exchange E-mail messages, and act as an electronic pager.

- *Wireless local area networks*—Examples of wireless local area networks include those such as Wireless Ethernet in which local area network workstations communicate with the network server via short- to medium-distance radio waves. Many businesses and universities are installing or are planning to install wireless local area networks in specific application areas. For example, a major airline is planning to install a wireless network that would enable ticket agents to meet the passengers as they exit their automobiles and check their luggage with the sky cap. This should eliminate the long lines passengers often encounter at the ticket counter and speed them toward their departing flights.

- *Wireless wide area networks*—Wireless wide area networks include a small but growing number of mobile wide area networks such as the RAM and ARDIS networks that operate using FM voice channels within the United States.

- *Satellite wide area networks*—Satellite wide area networks are systems that allow satellite communication to sparsely populated locations within the United States and eventually around the globe. A number of telephone and electronic companies are planning wide area satellite-based telephone/radio networks that would span the continent and eventually the world. Motorola has an especially ambitious project for spanning the entire globe with a network of satellites. First estimates were that 77 satellites would be necessary to completely cover the earth with telephone service. Since the element iridium has 77 electrons surrounding a nucleus, the project was titled Iridium. Since that time, the number of satellites has been reduced to 66. The concept of a single personal telephone number that would follow a person anywhere in the world is slowly coming closer to reality.

■ *Personal communications networks*—More prevalent in Europe, personal communication networks allow individuals to carry radio frequency units that are routed and controlled by local telephone exchange networks.

To provide many of the services described above, new transmission and modulation techniques have been and will be incorporated. Techniques that were once specialized only for the government, such as **spread spectrum** transmission, will continue to move into the commercial market. One of the more popular forms of spread spectrum transmission is **frequency hopping**. With this technique, multiple frequencies represent different data bits. As the data bits are transmitted, the carrier frequency changes with each transmitted bit. The intended receiver must follow this frequency hopping in order to receive the transmission properly, but the technique eliminates eavesdropping by intruders.

Another new transmission technique developed for wireless transmission is **code division multiple access** (**CDMA**). Similar to time division multiple access (TDMA), CDMA breaks each transmission stream into packets and assigns a unique code to each. All coded packets are then combined mathematically into one signal. Each intended receiver extracts only its data packets depending upon the assigned code.

A third new popular technology that has the potential for widespread use is **cellular digital packet data** (**CDPD**). Cellular digital packet data involves the transmission of data in the 1.90 to 1.92 GHz band, right among the frequencies used for voice cellular radio. Because of this location, companies that design and produce voice cellular radio (telephone) units will also be able to design and produce units capable of transmitting digital data over cellular networks. CDPD will open the market to an explosion of devices that will be able to transmit computer data to a remote location using the increasingly vast cellular network.

While these are some of the more common examples of new and emerging technologies that support wireless systems, there are many others, and many more will be created in the not-so-distant future.

INFORMATION SUPERHIGHWAY

Another very broad service access that has been much heralded but is still not here is the information superhighway. A term that apparently originated with Al Gore during the 1992 presidential campaign, the *information superhighway* (or more recently the National Information Infrastructure) was meant to describe a futuristic wide area network that would exist within the United States and provide a multitude of electronic services and information for homes, schools, businesses, and the government. To provide the power and flexibility needed for such services, some form of joint venture between telephone companies and cable television companies was thought to be necessary. While mergers have been attempted, government regulations and business economics have so far prevented any from being completed.

Among the application areas that would be available from an information superhighway are the following:

■ Online access to libraries, but more than just access to card catalogs: access to the books themselves

- Home shopping and banking transactions
- Ability to perform work duties from home
- Instant access to movies and interactive video
- Access to most national and international wide area networks, including electronic mail facilities, remote login, and file transfers
- Access to college campuses, including home study courses
- Interactive sessions with the United States Congress and legislators
- Online access to medical, legal, and other professional services

Many specialists in the field consider a single "superhighway" extremely unlikely and very limiting. Many envision a number of companies offering various services to homes and businesses. Most specialists agree that we are still a number of years away from any form of interactive information network for the personal user of the magnitude envisioned in the early 1990s. Unfortunately, it does not stop people and the press from using the term *information superhighway* to describe anything and everything that deals with dial-up information services.

SUMMARY

Trying to stay abreast of a quickly changing technology is indeed a difficult task. The field of study involving computers has evolved rapidly since their introduction in the 1950s, and many believe that the field of computer communications has evolved even more quickly than that of computers themselves. Considering the speed with which changes are occurring in this field we acknowledge that it is indeed impossible to learn everything by reading one book. We hope that through this text you will gain sufficient knowledge of the essentials to be able to apply it to numerous application areas and further studies. With these essentials, you may soon find yourself a participant in the exciting field of data communications and computer networks.

APPENDIX: FURTHER READINGS*

Chapter 1: Introduction

AT&T Breakup

A Slippery Slope: The Long Road to the Breakup of AT&T, Fred W. Henck. Greenwood Press, New York: 1988.

Teleprocessing

Introduction to Teleprocessing, James Martin. Prentice Hall, Englewood Cliffs, NJ: 1972.

OSI and Internet Models

Open Networking with OSI, Adrian Tang and Sophia Scoggins. Prentice Hall, Englewood Cliffs, NJ: 1992.

Computer Networks, 2nd Ed., Andrew S. Tanenbaum. Prentice Hall, Englewood Cliffs, NJ: 1988.

OSI: A Model for Computer Communications Standards, Uyless Black, Prentice Hall, Englewood Cliffs, NJ: 1991.

The Simple Book: An Introduction to Management of TCP/IP-Based Internets, Marshall T. Rose. Prentice Hall, Englewood Cliffs, NJ: 1991.

Open Systems Networking: TCP/IP and OSI, David M. Piscitello and A. Lyman Chapin. Addison Wesley, Reading, MA: 1993.

---

* The sources listed here have been limited primarily to textbooks to allow easier access to the average college student.

Chapter 2: Physical Layer (Part One)

Basic Concepts

Data Communications: A User's Guide, 3rd Ed., Ken Sherman. Prentice Hall, Englewood Cliffs, NJ: 1990.

Data Networks: Concepts, Theory, and Practice, Uyless Black. Prentice Hall, Englewood Cliffs, NJ: 1989.

Chapter 3: Physical Layer (Part Two)

Interface Standards

Data Networks: Concepts, Theory, and Practice, Uyless Black. Prentice Hall, Englewood Cliffs, NJ: 1989.

Data and Computer Communications, 4th. Ed., William Stallings. Macmillan, New York: 1994.

Physical Level Interfaces and Protocols, Uyless Black. Computer Society Press, Washington, D.C.: 1988.

EIA232D

Technical Aspects of Data Communication, 3rd Ed., John E. McNamara. Digital Press, Rockport, MA: 1988.

Modems

The Complete Modem Reference, Gilbert Held. Wiley, New York: 1991.

All About Modems, John A. Jurenko. Universal Data Systems, Huntsville, AL: 1981.

Chapter 4: Data Link Layer

BISYNC, SDLC, HDLC

Technical Aspects of Data Communication, 3rd Ed., John E. McNamara. Digital Press, Rockport, MA: 1988.

IBM Synchronous Data Link Control—General Information, IBM Manual no. GA27-3093-2. Research Triangle Park, NC: 1979.

IBM Binary Synchronous Communications—General Information, IBM Manual no. GA27-3004-2. Research Triangle Park, NC: 1970.

Protocols and Techniques for Data Communications Networks, Franklin F. Kuo. Prentice Hall, Englewood Cliffs, NJ: 1981.

Chapter 5: Introduction to Networks

Routing

Computer Networks, 2nd Ed., Andrew S. Tanenbaum. Prentice Hall, Englewood Cliffs, NJ: 1989.

Chapter 6: Local Area Networks

LANS

Handbook of Computer-Communications Standards: Volume 2 Local Network Standards, William Stallings. H. W. Sams. Indianapolis: 1987.

Local Networks, 3rd Ed., William Stallings. Macmillan, New York: 1990.

Local Area Networks: An Introduction to the Technology, John E. McNamara. Digital Press, Burlington, MA: 1985.

Bridges and Routers

Internetworking and Addressing, Gene White. McGraw-Hill, New York: 1992.

IEEE 802

Handbook of Computer-Communications Standards: Volume 2 Local Network Standards, William Stallings. H. W. Sams. Indianapolis: 1987.

Internetworking and Addressing, Gene White. McGraw-Hill, New York: 1992.

Chapter 7: Wide Area Networks

X.25

Internetworking and Addressing, Gene White. McGraw-Hill, New York: 1992.

The X Series Recommendations: Protocols for Data Communications Networks, Uyless Black. McGraw-Hill, New York: 1991.

TCP/IP

Internetworking with TCP/IP: Principles, Protocols, and Architecture, Volume 1, 2nd Ed., Douglas E. Comer. Prentice Hall, Englewood Cliffs, NJ: 1991.

The Simple Book: An Introduction to Management of TCP/IP-Based Internets, Marshall T. Rose. Prentice Hall, Englewood Cliffs, NJ: 1991.

Open Systems Networking: TCP/IP and OSI, David M. Piscitello and A. Lyman Chapin. Addison Wesley, Reading, MA: 1993.

TCP/IP Illustrated: Volume 1 The Protocols, W. Richard Stevens. Addison Wesley, Reading, MA: 1994.

SNA

Introduction to SNA Networking Using VTAM/NCP, Jay Ranade and George C. Sackett. McGraw-Hill, New York: 1989.

ISDN, DECnet and SNA Communications, Thomas C. Bartee. H. W. Sams. Indianapolis: 1989.

SNA: IBM's Networking Solution, James Martin. Prentice Hall, Englewood Cliffs, NJ: 1991.

Systems Network Architecture, Concepts and Products, IBM Manual no. GC30-3072-1. Research Triangle Park, NC: 1981.

Chapter 8: Satellite and Radio Networks

Basic Concepts

Computer Communications: Volume 1 Principles, Wushow Chou, Ed. Prentice Hall, Englewood Cliffs, NJ: 1983.

Data and Computer Communications, 3rd Ed., William Stallings. Macmillan, New York: 1988.

Chapter 9: Transport and Session Layers

Basic Concepts

Computer Networks, 2nd Ed., Andrew S. Tanenbaum. Prentice Hall, Englewood Cliffs, NJ: 1988.

Client-server

Internetworking with TCP/IP, Volume III: Client-Server Programming and Applications for the BSD Socket Version, Douglas E. Comer and David L. Stevens. Prentice Hall, Englewood Cliffs, NJ: 1993.

Chapter 10: Presentation and Application Layers

ASN.1

The Open Book: A Practical Perspective on OSI, Marshall T. Rose. Prentice Hall, Englewood Cliffs, NJ: 1990.

Compression

Data Compression Techniques and Applications, Thomas J. Lynch. Lifetime Learning Publications, Belmont, CA: 1985.

Encryption

Top Secret: Data Encryption Techniques, Gilbert Held. H. W. Sams, Carmel, IN: 1993.

Computer Networks, 2nd Ed., Andrew S. Tanenbaum. Prentice Hall, Englewood Cliffs, NJ: 1988.

Chapter 11: Emerging Technologies

ISDN, Broadband ISDN, ATM

Network Standards: A Guide to OSI, ISDN, LAN, and MAN Standards, William Stallings. Addison Wesley, Reading, MA: 1993.

Fiber Optics

Fiber Optic Communications, Harold B. Killen. Prentice Hall, Englewood Cliffs, NJ: 1991.

Information Highway

The Traveler's Guide to the Information Highway, Dylan Tweney. Ziff-Davis Press, Emeryville, CA: 1994.

GLOSSARY OF ACRONYMS AND ABBREVIATIONS

4B/5B code   Ecoding scheme used for fiber optic systems. The first step is to convert 4-bit quantities of the original data into new 5-bit quantities.

A   Bit that is part of the frame status (FS) field of the IEEE 802.5 token ring protocol packet format; indicates whether the receiving station recognizes the packet address.

(AC) access control   Together with the frame control field provides priority information and other control information; part of the IEEE 802.5 token ring protocol packet format.

(ACK0) ACKnowledgment 0   BISYNC control code indicating positive acknowledgment to even-sequenced frames of data or as a positive response to a select or bid (*see* BSC, ACK1).

(ACK1) ACKnowledgment 1   BISYNC control code meaning positive acknowledgment to odd-sequenced frames of data (*see* BSC, ACK0).

(ADCCP) Advanced Data Communication Control Protocol   Bit-synchronous protocol developed by the American National Standards Institute.

Alohanet   Data transfer system used in Hawaii; centralized packet radio network that allows remote terminals wireless access to a mainframe.

(ANSI) American National Standards Institute   Primary organization involved in deciding the standards used in interfaces to ensure that equipment manufactured by one company can be used in conjunction with that made by another.

ARDIS   Mobile wide area network that uses FM voice channels within the United States.

ARPANET   Wide area network designed by the U.S. government as a research vehicle in the late 1960s. It connected select research universities, military bases, and government labs across the United States. It was split in 1983 into MILNET for military use and ARPANET for experimental research. It has evolved into the present-day Internet.

(ASCII) American National Standard Code for Information Interchange Data code or system in which each character is represented by a unique pattern of 1s and 0s. ASCII is a 7-bit code, meaning that each character has its own configuration of seven 1s and 0s. ASCII has 128 possible combinations, including uppercase and lowercase letters, the digits 0 to 9, special symbols, and control characters (*see* EBCDIC).

(ASN.1) abstract syntax notation Protocol used for transmission of complex records and data structures from one network application to another. It consists of two parts: the description language, a Pascal-like code used to define the data structure; and the transfer syntax notation used to transmit the data structure. It is part of the data representation service of the presentation layer of the Open Systems Interconnection (OSI) model.

(AT&T) American Telephone and Telegraph Company Telephone company that handles long-distance calls. A call from one local access and transport area (LATA) to another would be handled by a long-distance company such as AT&T.

(ATM) asynchronous transfer mode Transmission technique chosen for use in broadband integrated services digital networks (ISDN). It allows the multiplexing of numerous logical connections over a single physical connection.

(ATM) automatic teller machine Machine operating 24 hours a day that can perform most routine banking operations—withdrawals, deposits, and transfers—from locations remote from the bank.

AX.25 Popular standard for distributed packet radio networks. It is similar to the high-level data link control protocol.

(BCC) Block Check Count Error checksum appended to the end of the data.

(BCN) Test Response/Beacon In the synchronous data link control (SDLC) protocol, one of the available unnumbered frame responses.

(BDLC) Burroughs Data Link Control Bit-synchronous protocol developed by Burroughs.

(BISDN) Broadband-ISDN Integrated services digital network (ISDN) that supports data rates over 600 Mbps. It was designed for use with fiber optic cable in order to handle that medium's high data rate (*see* ISDN).

BITNET Educational electronic mail system.

(BOC) Bell Operating Company Part of AT&T before the court decision of 1984 that required AT&T to divest itself of companies that provided local telephone service. Before 1984, AT&T consisted of 23 BOCs providing local service across the country.

(bps) bits per second A measure of the speed at which data can be transferred electronically (*see* Mbps and Gbps).

(BSC) BISYNC Protocol designed by IBM to provide a general-purpose data link mechanism for point-to-point and multipoint connections. It allows data to be transmitted in both directions, but not at the same time. It is dependent on a data code, such as ASCII or EBCDIC.

C Bit that is part of the frame status (FS) field of the IEEE 802.5 token ring protocol packet format; indicates whether the receiving station copies (accepts) the packet.

(CCITT) Comite Consultatif International Telephonique et Telegraphique Primary organization involved in deciding the standards used in interfaces to ensure that equipment manufactured by one company can be used in conjunction with that made by another.

CCITT V.28   Standard on which the electrical, functional, and procedural components of the EIA-232-D interface are based. It describes the electrical characteristics for an unbalanced interchange circuit and defines a list of 43 interchange circuits that can be used in an interface design.

(CDDI) copper distributed data interface   Variation of the fiber-distributed data interface (FDDI); has a format similar to FDDI, but uses unshielded twisted pair cable in place of fiber optic cable.

(CDMA) code division multiple access   Technique that breaks each transmission into packets and assigns a unique code to each. All coded packets are then combined mathematically into one signal and each intended receiver extracts only its data packets depending on the assigned code.

(CDPD) cellular digital packet data   Transmission technique that transmits data in the 1.90 to 1.92 GHz band.

(CFGR) Configure for Test   One of the available unnumbered frame commands in the synchronous data link control protocol.

(codec) COder/DECoder   Device that is essentially the opposite of a modem; it inputs analog data and produces digital signals.

(CR) carriage return   Command telling a printer to move to the left margin on the current print line.

(CRC) cyclic redundancy checksum   Type of error detection that treats a packet of data for transmission as a large polynomial. It uses fewer check digits and obtains greater accuracy than other error detection mechanisms.

(CR-LF) carriage return-line feed   Character-code sequence used in the network virtual terminal (NVT) format to mark the end-of-line transmission sequence.

(CSMA) carrier sense multiple access   Packet radio access protocol. It works well when stations are relatively close together; however, it cannot detect when a collision of data has occurred.

(CSMA/CD) carrier sense multiple access with collision detection   Access protocol that allows the workstation to "listen" to the medium (to "sense" for a carrier) to learn whether anyone else is transmitting data; if another transmission is heard, the assumption is that a collision has occurred.

(DCE) data circuit-terminating equipment   Standardized nomenclature referring to devices such as modems and multiplexors.

(DDCMP) Digital Data Communication Message Protocol   Bit-synchronous protocol developed by the Digital Equipment Corporation.

(DISC) DISConnect   One of the available unnumbered frame commands in the synchronous data lick control protocol.

(DLE) data link escape   BISYNC control code inserted before a control character to ensure that that character is read as a control; prevents regular text characters similar to control characters from being misread as controls (*see* BSC).

(DM) Request Online   In the synchronous data link control (SDLC) protocol, one of the available unnumbered frame responses.

(DOV) data over voice   Hardware associated with some PBXs that allows computer workstations and voice telephones to share a single line for both data and voice transmissions.

(DTE) data terminal equipment   Standardized nomenclature referring to devices such as terminals and computers.

(E) error detected bit   Part of the ending delimiter of the IEEE 802.5 token ring protocol packet format.

(EBCDIC) Extended Binary Coded Decimal Interchange Code   Data code or system in which each character is represented by a unique pattern of 1s and 0s. EBCDIC is an 8-bit code, meaning that each character has its own configuration of eight 1s and 0s. This code has 256 possible combinations including all uppercase and lowercase letters, the 10 digits, a large number of special symbols and punctuation marks, and a number of control characters (*see* ASCII).

(ED) ending delimiter   Part of the IEEE 802.5 token ring protocol packet format.

(EIA) Electronics Industries Association   Primary organization involved in deciding the standards used in interfaces to ensure that equipment manufactured by one company can be used in conjunction with that made by another.

EIA-232-D   Interface standard for connecting data terminals to voice-grade modems for use on analog public telecommunications systems using synchronous or asynchronous transmission. This replaced the older RS-232c standard (*see* RS-232c). The EIA-232-D defines all four components of an interface standard: mechanical interface, electrical, functional, and procedural.

(E-mail) electronic mail   Communications application that allows one user in a computer network to send a message directly to the computer of another user in the network.

(END DELIM) ending delimiter   Part of the medium access control sublayer packet format for IEEE 802.3 CDMA/DC.

(ENQ) Inquiry   BISYNC control code used to initiate a poll message or select message, to bid for the line in contention mode, or to ask that a response to a previous transmission be resent (*see* BSC).

(EOT) End Of Transmission   BISYNC control code used to signal the end of a transmission (*see* BSC).

(ETB) End of Transmission Block   BISYNC control code used to signal the end of transmission of a block of data (*see* BSC).

(ETX) End of TeXt   BISYNC control code used to signal the end of a complete text message (*see* BSC).

(FC) frame control   Together with the access control field provides priority information and other control information; part of the IEEE 802.5 token ring protocol packet format.

(FCC) Federal Communications Commission   Government entity that controls the allocation of available airwave frequencies within the United States.

(FCS) Frame Check Sequence   Field included to detect errors in the data block used in statistical time division multiplexing.

(FDDI) fiber-distributed data interface   Protocol that resembles a token ring network that has been enhanced—with higher data transmission speeds, longer transmission distances, and more possible interconnections.

(FDM) frequency division multiplexing   Assignment of different, nonoverlapping frequencies to users of the same medium, such as to the multiple users of the coaxial cable that delivers cable television.

(FDMA) frequency division multiple access   User access to a satellite system controlled through assignment of a group of frequencies to each user. A disadvantage is that frequencies are wasted acting as buffer regions to prevent the assigned channels from overlapping other frequencies.

(FRMR) Command Reject   One of the available unnumbered frame responses in the synchronous data link control protocol; indicates that the receiver rejects the frame or can't make sense of it.

(FS) frame status field   Final field of the IEEE 802.5 token ring protocol packet format (*see* A, C, and R).

(FTAM) File Transfer, Access, and Management   Attempt by the International Standards Organization (ISO) to create a generic file transfer standard for an OSI-type network.

(FTP) File Transfer (Protocol)   Part of the Internet protocol used for transferring files from one computer system to another.

(Gbps) gigabits per second   1 billion bits per second; one measure of the speed at which data can be transferred electronically (*see* bps and Mbps).

(GHz) gigahertz   1 billion cycles per second; one measure of the speed at which radio waves can be transmitted (*see* Hz, Khz, and Mhz).

H0   384-Kbps data channel capable of carrying high-speed data for applications such as facsimile, video, high-quality audio, or multiple lower-speed data streams.

H11   1.536-Mbps data channel capable of carrying high-speed data for applications such as facsimile, video, high-quality audio, or multiple lower-speed data streams.

H12   1.92-Mbps data channel capable of carrying high-speed data for applications such as facsimile, video, high-quality audio or multiple lower-speed data streams.

(HDLC) High-level Data Link Control   Data link standard created by the International Standards Organization (ISO) that closely resembles the synchronous data link control (SDLC) protocol.

(Hz) hertz   Unit of frequency equal to 1 cycle per second; used to measure the speed at which radio waves can be transmitted.

(I) intermediate frame bit   Part of the ending delimiter of the IEEE 802.5 token ring protocol packet format; indicates whether this is the last or only frame of transmission or other frames are to follow.

IBM 327x   Interface standard created by IBM to connect their 3270 family of terminals to other IBM equipment.

(ICMP) Internet Control Message Protocol   Software protocol designed to handle error and control messages. All gateways using this protocol have this software.

(IEEE) Institute for Electrical and Electronics Engineers   Primary organization involved in deciding the standards used in interfaces to ensure that equipment manufactured by one company can be used in conjunction with that of another company.

IEEE 802   Local area network protocol developed by the IEEE Computer Society Local Network Standards Committee.

IEEE 802.3   IEEE protocol standard for CSMA/CD; it has several different versions, depending on the transmitting medium (coaxial cable, unshielded twisted pair, etc.) (*see* IEEE, IEEE 802).

IEEE 802.4   IEEE protocol standard for a token bus; it has several different versions, depending on the transmitting medium (coaxial cable, unshielded twisted pair, etc.)

(*see* IEEE; IEEE 802).

IEEE 802.5   IEEE standard for token ring local area networks, simpler than either 802.3 or 802.4; popularized by IBM (*see* IEEE; IEEE 802).

Internet   Network established by the National Science Foundation's Division of Network Communications Research and Infrastructure with the original intention of encouraging the exchange of scientific and engineering research information.

(IP) Internet Protocol   Standard for a connectionless packet delivery service for transferring data from one computer network to another (*see* TCP; TCP/IP).

(IPM) interpersonal messaging   Part of the message handling system (MHS) in an electronic mail system that is responsible for the contents of the "envelope," or the message itself.

(ISDN) Integrated Services Digital Network   Worldwide public telecommunication network designed in the early 1980s to support telephone signals, data, and a multitude of other services for both residential and business users.

(ISO) International Standards Organization   Primary organization involved in deciding the standards used in interfaces to ensure that equipment manufactured by one company can be used in conjunction with that of another company.

ISO 2110   Mechanical interface standard used by the EIA-232-D interface standard that defines the size and configuration of a 25-pin connector of the mechanical interface.

(ITB) end of Intermediate Transmission Block   BISYNC control code used to signal the end of an intermediate transmission block (*see* BSC).

Kermit   File transfer program that was designed to transfer files between microcomputers and mainframes.

(KHz) kilohertz   1000 cycles per second; one measure of the speed at which radio waves can be transmitted (*see* Hz, GHz, and MHz).

(LAN) local area network   Communication network that interconnects a variety of data communicating devices within a small area such as a room, building, or group of buildings and transfers data at high transfer rates with very low error rates.

(LAP) Link Access Protocol   Modified high-level data link control (HDLC) bit-synchronous protocol developed by Comite Consultatif International Telephonique et Telegraphique.

(LAPB) Link Access Protocol B   Modified bit-synchronous protocol developed by Comite Consultatif International Telephonique et Telegraphique.

(LATA) Local Access and Transport Area   Geographical area assigned to a local telephone company. Calls within this area are handled by the local company and are considered local calls.

(LED) light-emitting diode   Light source used with fiber optic cable.

(LF) line feed   Command telling a printer to move to the next print line while remaining in the current column position.

(LLC) logical link control   Sublayer of the data link layer of a local area network, part of the IEEE Computer Society Local Network Standards for the interconnection of computers and associated peripherals in a local area (*see* IEEE 802).

(LU) logical unit   Category of a network addressable unit (NAU). The LU is the software that allows a user access to the Systems Network Architecture (SNA). Two end users (humans, terminals, software programs, etc.) can communicate by establishing an LU to LU session.

(MAC) medium access control   Sublayer of the data link layer of a local area network, part of the IEEE Computer Society Local Network Standards for the interconnection of computers and associated peripherals in a local area (*see* PRE, SD, and END DELIM).

(MAN) metropolitan area networks   Networks capable of serving a wide number of users over a large area such as a town or city.

(MAP) Manufacturing Automation Protocol   Token bus local area network created by General Motors for use on an automotive assembly line.

(MAU) multistation access unit   Mechanism by which each workstation attaches to a token ring local area network that follows the IEEE 802.5 standard, creating a star-wired ring configuration.

(Mbps) megabits per second   Measure of the speed at which data can be transferred electronically (*see* bps and Gbps).

MCI   Telephone company that handles long-distance calls (*see* AT&T).

(MHS) message handling systems   Attempt to standardize electronic mail. It defines a collection of services between the two end users of an electronic mail system.

(MHz) megahertz   1 million cycles per second; one measure of the speed at which radio waves can be transmitted (*see* Hz, GHz, and KHz).

MILNET   Main computer network used by the military. It was established in 1983 when ARPANET was split into two networks: MILNET for the military and ARPANET for experimental research (*see* ARPANET).

(MIN) mobile ID number   Built-in, unique, identification code of a mobile phone, used to keep track of a user's account for billing purposes.

(MNP) Microcom Networking Protocol   Standards developed by Microcom Systems, Inc., to provide error correction and data compression techniques for asynchronous modem data transfer. Both ends of a connection must use modems with the same MNP standards to be able to use the error correction and data compression techniques.

(modem) MOdulator/DEModulator   Device that converts digital data onto an analog signal and then reconverts the analog signal back to digital data. This conversion is necessary for sending data from a computer, which uses a digital signal, over a telephone line, which uses an analog signal.

MOTIS   International Standards Organization's (ISO) version of the message handling system (MHS) standards for electronic mail.

(MTS) message transfer system   Part of the message handling system (MHS) in an electronic mail system that is responsible for the "envelope" or the details surrounding the message.

(MTSO) mobile telephone switching office   Connection to the hardwire phone system that serves as a receiver of data from the cell transmitter in a cellular phone system.

(NAK) Negative AcKnowledgment   BISYNC control code used to indicate that the previously received block was in error. Also used as a negative response to a poll message (*see* BSC).

(NAU) network addressable unit   Piece of software that allows a user, an application, or a hardware device to access a network.

(NFS) Network File System   Part of the Internet Protocol designed by Sun Microsystems to allow access to UNIX file systems.

(NIC) network information center   Central authority that assigns the network portion of Internet addresses.

N(R) [Number (received)]   In the high-level data link control (HDLC) and synchronous data link control (SDLC) standards, a field that shows the number of frames received from a particular station.

(NRZ-I) Non-Return to Zero, Invert on ones   Format for encoding digital data that uses a voltage change at the beginning of a 1 and no voltage change at the beginning of a 0 (*see* NRZ-L and RZ).

(NRZ-L) Non-Return to Zero-Level   Format for encoding digital data. A logic 0 in the data is represented as a "high" voltage signal; a logic 1 in the data is represented as a "low" voltage signal (*see* NRZ-I and RZ).

N(S) [Number (sent)]   In the high-level data link control (HDLC) and synchronous data link control (SDLC) standards, a field that shows the number of frames sent from a particular station.

(NSFNET) National Science Foundation NETwork   Nationwide network commonly known as Internet (*see* ARPANET).

(NSI) NonSequenced Information   In the synchronous data link control (SDLC) protocol, one of the available unnumbered frame commands.

(NVT) network virtual terminal   Format in the TCP/IP suite of protocols by which all transmitted keystrokes are converted to a common form to allow different types of client terminals to communicate with the TELNET server.

(ORP) Optional Response Poll   In the synchronous data link control (SDLC) protocol, one of the available unnumbered frame commands.

(OSI) Open Systems Interconnection Model   Seven-layer model for a computer network delineating the functions that must be performed at each layer for the network to transfer information successfully to computers within the network.

(PAD) packet assembler/disassembler   Device that accepts low-speed asynchronous data from a terminal and transforms it into the packet format necessary to traverse the public data network.

(PBX) private branch exchange   Privately owned telephone switching system that many businesses and colleges use to provide all their in-house telephone services.

(PCM) pulse code modulation   Technique of converting analog data to digital signals by taking "snapshots" of the analog data at fixed intervals.

(PCN) personal communications networks   Category of wireless network systems.

(PDN) public data networks   Network that has the responsibility of transmitting the data from the network host data circuit-terminating equipment (*see* DCE) across the subnet to the destination DCE.

(P/F) Poll/Final   Indicator of end of field. When a primary is polling a secondary, the last message will have P/F bit set to 1 to as the final bit in a synchronous data link control protocol.

P(R)   In a public data network, same as N(R).

(PRE) preamble field   First part of the medium access control sublayer packet format for IEEE 802.3 CSMA/DC.

P(S)   In a public data network, same as N(S).

(PU) physical unit   Classification of a network addressable unit. An administrative-type program that allows the SNA network to bring a node online, perform tests on the node, and perform similar network management functions.

(QOS) quality of service   Parameters used by Class 2 protocols that specify preferred, acceptable, and unacceptable values for a number of events such as transit delay times, residual error rates, and throughput rates to determine whether a transport connection can be established successfully.

R   Bit that is part of the frame status (FS) field of the IEEE 802.5 token ring protocol packet format; these bits are reserved for future use.

RAM   Wireless wide area network that operates using FM voice channels within the United States.

(RBOCs) Regional Bell Operating Companies   New configuration adopted by the 23 BOCs after the 1984 court-ordered divestiture of local companies by AT&T. To survive, the BOCs regrouped into seven RBOCs to continue offering local telephone service (*see* BOC).

(RD) Disconnect Request   One of the available unnumbered frame responses in the synchronous data link control protocol.

(REJ) Reject   Supervisory message used in SDLC meaning negative acknowledgment; go back to the Nth frame and resend all frames from the Nth frame on (*see* SLDC).

(RIM) Request for Initialization   In the synchronous data link control (SDLC) protocol, one of the available unnumbered frame responses.

(RNR) Receive Not Ready   Supervisory message used in SDLC meaning positive acknowledgment, but not ready to receive information frames (*see* SDLC).

(RR) Receive Ready   Supervisory message used in SDLC meaning positive acknowledgment; ready to receive an information frame (*see* SDLC).

RS-232c   Early standard interface used to connect a data terminal to voice-grade modems for use on analog public telecommunications systems.

RS-422A (X.27) and RX-423A (X.26)   Standards that address only the electrical components of the interface standard RS-449.

RS-449   Intermediate interface introduced after the RS-232c and before the EIA-232-D.

(RVI) ReVerse Interrupt   BISYNC control code issued by the receiver of information when it needs to interrupt the sender in order to seize the line to transmit a high-priority message (*see* BSC).

(RZ) Return to Zero   Format for encoding digital data. This format uses a low to high signal transition at the beginning of a logic 1 and a high to low signal transition in the middle of the bit. The logic 0 has no signal transitions (*see* NRZ-L and NRZ-I).

(SABM) Set Asynchronous Balanced Mode   Numbered frames command in the high-level data link control (HDLC) protocol; peer-to-peer connection in which either station can initiate transmission without explicit permission from the other station.

(SABME) Set Asynchronous Balanced Mode Extended   Numbered frames command in the high-level data link control (HDLC) protocol; same as SABM but in extended mode (*see* SABM).

(SARM) Set Asynchronous Response Mode   Numbered frames command in the high-level data link control (HDLC) protocol; allows secondary to initiate transmission without explicit permission of the primary, but primary still retains responsibility for the line (initialization, error recovery, and logical disconnection).

(SARME) Set Asynchronous Response Mode Extended   Numbered frames command in the high-level data link control (HDLC) protocol; same as SARM but in extended mode (16-bit control field).

(SD) starting delimiter   First part of the IEEE 802.5 token ring protocol packet format.

(SD) starting delimiter   Part of the medium access control sublayer packet format for IEEE 802.3 CSMA/DC.

(SDDI) shielded twisted pair distributed data interface   Variation of the fiber-distributed data interface (FDDI); has a format similar to FDDI, but uses shielded twisted pair cable in place of fiber optic cable.

(SDH) Synchronous Digital Hierarchy   Network standardized by CCITT that defines a synchronous multiplexing format for transmitting low-level digital signals over fiber optic cables.

(SDLC) Synchronous Data Link Control   Bit-synchronous protocol developed by IBM in the mid-1970s to replace BISYNC. It does not rely on character codes to control execution and allows data to be transmitted in both directions at the same time.

(SIM) Set Initialization Mode   In the synchronous data link control (SDLC) protocol, one of the available unnumbered frame commands.

(SMTP) Simple Mail Transfer Protocol   Part of the Internet protocol that allows users to send and receive electronic mail.

(SNA) Systems Network Architecture   Wide area network designed by IBM. It is a model that describes how communications/network software should be created.

(SNMP) Simple Network Management Protocol   Part of the Internet Protocol that allows the many forms of computer networks within the Internet to be viewed as a simple network.

(SNRM) Set Normal Response Mode   In the synchronous data link control (SDLC) protocol, one of the available unnumbered frame commands.

(SNRME) Set Normal Response Mode Extended   Numbered frames command in the high-level data link control (HDLC) protocol; similar to SDLC but control field is 16 bits in length.

(SOH) Start Of Header   BISYNC control code used to indicate an optional header field follows immediately (*see* BSC).

(SONET) Synchronous Optical NETwork   Network proposed by BellCore and standardized by ANSI that employs fiber optic cables. It defines a synchronous multiplexing format for transmitting low-level digital signals.

(SREJ) Selective REJect   Supervisory frame used in HDLC to tell the sender that a message was in error and to go back to that one message and retransmit it but not all the messages that followed it.

(SSCP) System Services Control Point   Classification of a network addressable unit. It resides in a host computer and has control over all cluster controllers, communications controllers, and terminals attached to that host.

(STX) Start of TeXt   BISYNC control code used to indicate the start of text and that the data follows immediately (*see* BSC).

(SYN) SYNchronization   BISYNC control code used to synchronize the incoming message with the receiver; precedes all message frames (*see* BSC).

(TCP) Transmission Control Protocol   Software transport service that, laid over an unreliable connectionless packet delivery service, results in a reliable network, protected from lost or duplicate packets.

(TCP/IP) Transmission Control Protocol/Internet Protocol Two sets of standards that, used together, create a reliable internetwork transport and delivery system.

(TDM) time division multiplexing Transmitting bits of information from different users of the same medium at the same time, in alternating fashion, usually 1 bit from user A, then 1 from user B, etc.

(TDMA) time division multiple access Method for allotting access to the range of acceptable frequencies assigned to a satellite, allowing each user a specific time block in which to use the satellite system.

TELNET Relatively simple remote terminal protocol that operates over the ICP/IP suite of protocols. It allows a user at a remote site to establish a Transmission Control Protocol (TCP) connection to a logic server at a second site.

(TEST) Request Task Response In the synchronous data link control (SDLC) protocol, one of the available unnumbered frame commands.

(TOP) Technical and Office Protocols Token bus local area network created by Boeing as a standard for office automation.

(TP) transport layer Layer of software that receives a packet of data and delivers it to the Internet practical layer.

(TTD) Temporary Text Delay BISYNC control code used by the sending station to indicate that it cannot send data immediately but does not wish to relinquish control of the line (*see* BSC).

(TTY) TeleTYpwriter interface RS-232-type interface that uses a multiple-pin connector and a multiple-wire twisted pair; requires voltage levels and signals following the specifications for the RS-232 standard interface; used to interface terminals to host computers.

(UA) Nonsequenced Acknowledgment In the synchronous data link control (SDLC) protocol, one of the available unnumbered frame responses.

(UI) Nonsequenced Information In the synchronous data link control (SDLC) protocol, one of the available unnumbered frame responses.

(UPS) United Parcel Service International package delivery system.

(V) code violation bit Part of the ending delimiter of the IEEE 802.5 token ring protocol packet format.

Videotex Information transmission using video.

(VSAT) very small aperture terminal Single-user earth station satellite system with its own ground station and a small antenna. The ground stations communicate with a mainframe computer via the satellite and a master station.

V-series Interface standard created by CCITT.

(WACK) Wait before transmit positive ACKnowledgment BISYNC control code used by a receiving station to acknowledge a previous message but also to say it cannot accept any more messages until further notice (*see* BSC).

X.3, X.28, and X.29 Protocols used in combination to interface a low-speed asynchronous terminal to a public data network.

X.25 Standard created in 1974 by CCITT to allow a user to connect its terminal to a public data network; sets a uniform approach for connecting to a network station. Use of X.25 involves a high-speed synchronous interface and requires considerable software and computing power.

(XID) Request Station ID In the synchronous data link control (SDLC) protocol, one of the available unnumbered frame commands.

X-series   Interface standard.

X-Windows   Set of protocols that defines a graphic window system for networks.

INDEX